Practical Biostatistics

Practical Biostatistics

A Step-by-Step Approach for Evidence-Based Medicine

Second Edition

MENDEL SUCHMACHER

Professor and Chairman of Clinical Immunology, Carlos Chagas
Institute of Medical Graduation, Rio de Janeiro, Rio de Janeiro State,
Brazil

MAURO GELLER

Full Professor of Clinical Genetics, Phacomatosis Department, Federal
University of Rio de Janeiro, Brazil

ACADEMIC PRESS

An imprint of Elsevier

Academic Press is an imprint of Elsevier
125 London Wall, London EC2Y 5AS, United Kingdom
525 B Street, Suite 1650, San Diego, CA 92101, United States
50 Hampshire Street, 5th Floor, Cambridge, MA 02139, United States
The Boulevard, Langford Lane, Kidlington, Oxford OX5 1GB, United Kingdom

Notices
Knowledge and best practice in this field are constantly changing. As new research and experience broaden
our understanding, changes in research methods, professional practices, or medical treatment may become
necessary.

Practitioners and researchers must always rely on their own experience and knowledge in evaluating and
using any information, methods, compounds, or experiments described herein. In using such information or
methods they should be mindful of their own safety and the safety of others, including parties for whom they
have a professional responsibility.

To the fullest extent of the law, neither the Publisher nor the authors, contributors, or editors, assume any
liability for any injury and/or damage to persons or property as a matter of products liability, negligence or
otherwise, or from any use or operation of any methods, products, instructions, or ideas contained in the
material herein.

British Library Cataloguing-in-Publication Data
A catalogue record for this book is available from the British Library

Library of Congress Cataloging-in-Publication Data
A catalog record for this book is available from the Library of Congress

ISBN: 978-0-323-90102-4

For Information on all Academic Press publications
visit our website at https://www.elsevier.com/books-and-journals

Publisher: Stacy Masucci
Acquisitions Editor: Rafael E. Teixeira
Editorial Project Manager: Tracy I. Tufaga
Production Project Manager: Omer Mukthar
Cover Designer: Miles Hitchen

Typeset by MPS Limited, Chennai, India

Contents

About the authors

Mendel Suchmacher

Professor and Chairman of Clinical Immunology, Carlos Chagas Institute of Medical Graduation

Mendel Suchmacher, MD, MSc, holds an MD degree from the Teresopolis University Medical School, an MSc in Management, Research, and Development in Pharmaceutical Industry from the Fundacao Instituto Osvaldo Cruz, is board-certified in Internal Medicine and Hematology-Hemotherapy, is a member of the American College of Physicians, is a member of clinical staff of Arthur de Siqueira Cavalcanti Hematology Institute (Hemorio), and is a technical coordinator of Souza Aguiar Hospital Blood Bank (Ministry of Health). He is a clinical staff and researcher of the Genodermatoses Sector of the Clinical Genetics Service at Santa Casa de Miscricordia do Rio de Janeiro. He holds teaching and research positions at prestigious Brazilian institutions: professor of Clinical Immunology at the Carlos Chagas Institute of Medical Graduation and research fellow at the Teresopolis University Medical School—UNIFESO, having published 17 papers, 6 book chapters, and 6 books.

Mauro Geller

Full Professor of Clinical Genetics, Phacomatosis Department, Federal University of Rio de Janeiro, Brazil

Mauro Geller, MD, PhD, holds an MD degree from the Teresopolis University Medical School, a PhD in Clinical Medicine from the Federal University of Rio de Janeiro, and a Post-doc in Immunogenetics from Harvard University. He has extensive experience in the field of clinical immunology, especially in the areas of clinical medicine, tumoral immunology, genetics, and immunodiagnostics. Dr. Geller is a founding member and current medical director of the Brazilian National Neurofibromatosis Center; a fellow of the American College of Physicians and of the Royal Society of Medicine; a member of the European Society of Gene Therapy; and a member of the Brazilian Societies of Immunology, Microbiology, and Genetics. He is board-certified in *Internal Medicine, Immunology, Allergy*, and *Public Health*. He also has extensive experience with research in the areas of immunology, microbiology, and genetics, as well as with clinical research, and has published 137 papers, 8 book chapters, and 4 books. He also serves as an ad hoc advisor to the Brazilian National Institute of Health (ANVISA) and is a member of the clinical staff of the Hospital Israelita Albert Einstein.

PART 1

The investigator's hypothesis

The objective of Part I is to detail the very beginning of the scientific process, that is, the elaboration of the first questions the investigator formulates.

CHAPTER 1

Investigator's hypothesis and expression of its corresponding outcome

Scientific investigations start with an investigator who: (1) observes a phenomenon, (2) formulates a question regarding some aspect of this phenomenon, (3) elaborates a hypothesis based on this question, either to explain the phenomenon or to establish some kind of correlation, (4) tests his/her hypothesis under controlled conditions, and (5) expresses his/her conclusion.

Even though the above detailed method must apply to scientific projects conceived in any field, some differences in the way of interpreting studies results and applying learned principles prevail among the various disciplines. For example, the study object in Medical Sciences does not behave in a deterministic manner, as in Exact Sciences. We could not, for instance, insert biological data from a patient in a mathematical formula in order to predict with a 100% certainty if this patient might respond to a certain medication and if it will be safe. Therefore it is necessary to apply tools which will be able to, at least, determine efficacy and safety probabilities regarding a drug, vaccine, examination, or medical procedure.

In Biostatistics, clinical impression is replaced by the more objective probabilistic mathematics, in which counted observations in a population are analyzed through statistical models suitable to investigator's hypothesis as well as to study type and design. Nevertheless, differently from the practice setting where we may state that a diagnosis is likely or a given therapy is prone to work, in the probabilistic setting, we must assume that any given correlation is a casual finding (null hypothesis—H_0) and that this casuality must be "pushed aside" until a minimum probabilistic value (generally 5%) is reached in order to accept that whatever was conclude is not casual (alternative hypothesis—H_1). For instance, we conclude that there is an association between propranolol and decreased arterial blood pressure, expressing it this way: there is a 95% probability that this association is not casual (H_1) and 5% probability that it is (H_0). An H_1 of 95% is generally accepted as sufficient to "push H_0 aside" (or, in more practical terms, to reject it). From a clinical standpoint, that would be enough to take the finding that propronolol lowers arterial blood pressure, as a scientific fact.

Practical Biostatistics
DOI: https://doi.org/10.1016/B978-0-323-90102-4.00002-3

All the above means that there is not 100% absolute certainty in Biostatistics and in Evidence-Based Medicine, but only a 5% probability that one is incorrect (consequently 95% that one is correct). In other words, the aim of most of the clinical research is to try to reject H_0 and to express its conclusions correspondingly.

In order to detail this type of approach even further, consider a table that crosses the findings of your study with absolute truth (Table 1.1).

- Situation A—Your positive finding coincides with absolute truth, that is, it is *not* casual. For example, you conclude that a tested bisphosphonate increased bone mineral density when bisphosphonates *effectively* increase bone mineral density.
- Situation B—Your positive finding does *not* coincide with absolute truth, that is, it is casual. This is a type I error (tolerance for this type of error is conventionally 5%). For example, you conclude that a tested antihistamine increased bone mineral density when antihistamines have *no* bone mineral density increasing effect.
- Situation C—Your negative finding does *not* coincide with absolute truth, that is, it is casual. This is a type II error (tolerance for this type of error is conventionally 20%). For example, you conclude that a tested bisphosphonate did *not* increase bone mineral density when bisphosphonates *effectively* increase bone mineral density.
- Situation D—Your negative finding coincides with absolute truth, that is, it is *not* casual. For example, you conclude that a tested antihistamine did *not* increase bone mineral density when antihistamines have *no* bone mineral increasing effect.

Normally, situation A is the one pursued in clinical research, that is, we wish to demonstrate that a medication or a vaccine works, not the opposite. Therefore we can express a conclusion by stating that P (type I error probability) inferred from a study is less, equal to or greater than α (statistical significance level which corresponds to the highest tolerable cutoff for type I error—often 0.05), as preestablished by the investigator. So, we would be authorized to reject H_0 and to accept H_1, once P (or P_α) $< \alpha$.

> *"We concluded that the trimethoprim-sulfamethoxazole combination was effective for Pneumocystis jiroveci lung infection control in the HIV-carrier sample studied (P < 0.05)."*

Table 1.1 The four possible situations derived from crossing an investigator's finding with the absolute truth.

		Absolute truth	
		+	−
Your finding	+	Situation A	Situation B
	−	Situation C	Situation D

Interpretation: there is a less than 5% probability that the above conclusion represents an error, that is, that the association between *Pneumocystis jiroveci* lung infection control in the HIV-carrier sample studied and therapy with trimethoprim-sulfamethoxazole combination, is a coincidence.

Therefore it is evident that the goal of clinical research is mostly to determine if $P < \alpha$. There are plenty of available mathematical tools meant to establish this correlation, which will be chosen by the biostatistician and the investigator as the fittest for the planned study design. Selecting them is a process involving many different elements and steps, to be detailed in the following chapters.

PART 2

Collective health

The objective of Part II is to detail the basics of collective health mathematics.

CHAPTER 2

Disease frequency measures

2.1 Preamble

Using multiples of 10 such as 100, 1000, 10,000, etc. as denominators for disease frequency measures (prevalence and incidence) is generally recommended. This normatization allows: (1) direct comparisons between different populations and (2) adjustments according to the rarity of a particular condition (e.g., using a denominator of 1,000,000 for a rare disease). Alternatively, disease frequency measures can be expressed with person-time units as the denominator instead of population size, for example, 0.05 cases/person-year.

2.2 Simple count

Simple count expresses the simple summation of cases of a condition in a given population. For example, in population A (P_A), there were 39 multiple myeloma cases out of a population of 934 monoclonal gammopathy of undetermined significance (MGUS) patients, and in population B (P_B), there were 54 multiple myeloma cases out of a population of 8344 MGUS patients. An obvious limitation of simple count is that it does not allow for direct comparisons between different populations, because simple summations performed in different contexts can have different meanings. In the above example, even though 54 cases sound more significant than 39 cases, those latter cases occurred in a population of 934 individuals and the former ones in a population of 8344 individuals.

2.3 Prevalence

Prevalence expresses the number of persons in a population who present a specific condition—new as well as preexisting cases—at a given point or period in time. This measure can be considered as a suitable tool for chronic conditions, for it considers preexisting cases as well. There are two types of prevalence: (1) point prevalence and (2) period prevalence.

2.3.1 Point prevalence

Point prevalence corresponds to the rate of cases of a given condition in a population at a specific point in time. It is determined by the formula:

$$PP = C_{ST}/N_{ST} \tag{2.1}$$

Practical Biostatistics
DOI: https://doi.org/10.1016/B978-0-323-90102-4.00017-5

PP is point prevalence, C_{ST} is number of cases of the illness at a specific point in time, and N_{ST} is number of individuals in the population at a specific point in time.

By applying the above example (Section 2.2), in P_A, 39 out of 934 MGUS patients have multiple myeloma. Therefore

$$PP_A = \frac{39}{934} = 0.041$$

In P_B, 54 out of 8344 MGUS patients have multiple myeloma. Therefore

$$PP_B = \frac{54}{8344} = 0.006$$

We can infer that P_A has a higher point prevalence of multiple myeloma than P_B. Another way of expressing the above results is:

$$PP_A = 41{:}1000 \ \text{MGUS} \ \text{patients}$$

$$PP_B = 6{:}1000 \ \text{MGUS} \ \text{patients}$$

Some authors use the terms point prevalence, prevalence, and morbidity rate interchangeably.

2.3.2 Period prevalence

Period prevalence corresponds to the rate of cases of a given condition in a population within a time interval. It is determined by the formula:

$$P_eP = C_t/N_m \tag{2.2}$$

P_eP is period prevalence, C_t is number of cases of the illness within the time interval, and N_m is number of individuals in the population at the midpoint of the time interval.

Presuming period prevalence determination involves a timespan, one must assume that the former might involve dynamic populations. So, N_m can be estimated by determining the mean between the number of individuals at the start of the time interval and the number of individuals at its end. For example, 39 multiple myeloma cases in MGUS patients were recorded during a 1-year period, out of a target population that counted 962 individuals at the study start and 1002 individuals at its end $[N_m = (962 + 1002)/2 = 982]$. Therefore

$$P_eP = \frac{39}{982} = 0.039 \ (\text{or} \ 39 : 1000)$$

The main advantage of period prevalence over point prevalence is that the former provides a broader picture, presuming period prevalence encompasses a timespan.

Therefore inferences drawn from comparisons between two different period prevalences can be more reliable than between two different point prevalences. Nevertheless, comparisons between two different period prevalences can only be made if respective timespans are identical and do not overlap. Choosing the best disease frequency measure will vary according to study objective and available resources.

2.4 Incidence

Incidence expresses the rate of incident cases (i.e., new cases) of a condition in a population at risk within a time interval. For determining incidence, one must assume that: (1) individuals at risk did not present the studied illness at the beginning of the considered timespan, (2) each individual is observed at least twice along the considered timespan for the occurrence of an incident, and (3) it will have been the first time the condition has compromised the individual within the considered timespan. Incidence is a suitable tool for studying epidemics and outbreaks, since it involves counting incident cases. Its main limitation is that it hardly takes into account some temporal aspects of the studied population, such as: (1) the exact time when a specific subject enter the studied population, (2) how long exactly did this subject remain in it, and (3) when exactly did the illness happen (if ever).

Incidence can be expressed in two ways:

2.4.1 Cumulative incidence

Cumulative incidence expresses the risk of contracting an illness in a given population within a timespan. It is determined by the formula:

$$C_{in} = IC/N_0 \tag{2.3}$$

C_{in} is cumulative incidence, IC is incident cases, and N_0 is number of healthy individuals at the beginning of the timespan.

Let us apply the example detailed in Section 2.2, assuming a 1 year timespan for P_A:

$$C_{in(t_0,t)} = \frac{39}{934} = 0.04$$

Cumulative incidence for P_A is 0.04 in 1 year. Another way of expressing it is that each individual woman had a 4% risk of representing a new case of multiple myeloma, during the 1 year timespan.

Cumulative incidence assumes that a fixed population is under study. But in real world situations, this is hardly the case. In most instances, cumulative incidence studies involve dynamic populations, meaning the existence of a Δt (time interval a particular individual spent in the study). Based on actuarial data, it is assumed that, if dropouts are meant to happen, they might do so at half the folow-up time. Therefore a second cumulative incidence formula can be applied:

$$C_{in} = (IC/N_0) - (W/2) \qquad (2.4)$$

C_{in} is cumulative incidence, IC is incident cases, N_0 is number of healthy individuals at the beginning of the timespan, and W is dropouts.

By applying the above example (presuming 80 patients might dropout):

$$C_{in} = \left(\frac{39}{934}\right) - \left(\frac{80}{2}\right) = 0.043$$

Some authors use the terms cumulative incidence and incidence proportion interchangeably, for the numerator (IC) represents a proportion of the denominator.

2.4.2 Incidence rate

Incidence rate expresses the occurrence rate of an illness in a given population. This disease frequency measure assumes that a dynamic population is under study, that is, either new individuals can be added to it or participating individuals can dropout (most often the latter situation). Therefore a sort of denominator, different than N_0, that takes temporal aspects into consideration must be applied: person-time. So, incident rate is determined by the formula:

$$IR = IC/PT \qquad (2.5)$$

IR is incidence rate, IC is incident cases, and PT is person-time (days, months, years).

Typically, each individual in the population is observed until one of four events occurs:
- Disease onset

    A person can be considered an incident case, therefore useful for determining incident rate. Nevertheless, assuming the former interrupted his or her participation during the study, then that person can only be partially counted concerning time. For example, if an individual participated in the study during 2 full years and the disease onset occurred at the eighth month of year 3, then that person will be counted as 2.8 persons-year.
- End of the study

    A person fulfilled the study timespan, then that individual can be fully counted as person-year. For example, if the study lasted 5 years, that individual will be counted as 5 persons-year.
- Death

    If a person died due to the studied illness, then that individual will be counted as an incident case.

- Dropout/lost to follow-up

    Even though it will obviously not be possible to count a person as an incident case, the former can nevertheless be included in the denominator of formula (2.5). As a rule of thumb, a person can be counted as 0.5 year, regarding the year of dropout/lost to follow-up occurrence. So, if an individual participated in the study along 2 full years and dropped out/was lost to follow-up during year 3, then the former can be counted as 2.5 persons-year.

    For example, 703 men in their sixth decade of life were followed annually along 5 years to determine the incidence rate of benign prostatic hyperplasia (BPH) (Fig. 2.1). After the first year, none had BPH. At the end of the second year, 12 men had BPH. At the end of the third year, 4 men had the condition, and at the end of the fourth year, 12 men.

    Notice that, once an incident case happened, then the respective patient stopped representing the population studied (men in their sixth decade of life) and was excluded from the study exactly at the incident case timepoint. Calculating: (1) $IC = 0 + 12 + 4 + 12 = 28$ and (2) $PT = 703$ individuals $\times$ 4 years $= 2812$ persons-year:

$$IR = \frac{28}{2812} = 0.009 \text{ cases/person-year}$$

Another way of expressing the above result is nine cases per 1000 persons-year.

Some authors use the terms incidence rate and morbidity rate interchangeably. Differences between cumulative incidence and incidence rate are detailed in Table 2.1.

2.5 Relationship between prevalence and incidence

Depending on the circumstances, both prevalence and incidence can be used to study the very same epidemiologic phenomenon. As a matter of fact, differences between both can

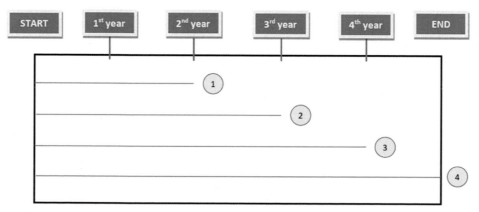

Figure 2.1 Diagram representing the evolution pattern of men in their sixth decade of life. At the end of the study, 675 individuals remained representing no BPH episodes.

Table 2.1 Some differences between cumulative incidence and incidence rate.

Cumulative incidence	Incidence rate
Dynamic populations can compromise accuracy	Dynamic populations do not compromise accuracy due to methodological flexibility provided by person-time parameter
Accepts well fixed populations	Accepts well fixed populations (even though it was not meant for it)
Suitable for short-time studies	Suitable for long-time studies
Disease risk is assumed as stable along the study	Person-time parameter can be biased due to varying degrees of disease risk, prevalent along the study[a]

[a]For example, 10 patients followed along one year for incident cases of myocardial infarction would equal one person followed for the same type of incident case along 10 years—when it is widely known the risk related to this type of disease can vary along a decade span.

sometimes be subtle, as if they represented two sides of the same coin. Nevertheless one can assume that, even though mutual influence is indeed expected to happen, prevalence, as the most "static" of the two, might be influenced rather than to influence. So, we can comment on three different factors that might influence prevalence:

- Incidence

    Incident cases will eventually become prevalent ones, therefore suitable as numerators for formulas (2.1) and (2.2). Similarly, prevalence is expected to vary directly and positively with incidence, that is, if incidence either increases or decreases, prevalence is expected to follow it.

- Disease duration

    Cure or death terminates the participation of an individual in prevalence counting. Nevertheless, the duration pattern of their previous condition can influence prevalence in two ways:

    - Short-duration conditions

        Short-duration conditions tend to yield lower prevalences. For example, stage IV pancreatic cancer, which rapidly evolves to death. Similarly, conditions with a high incidence, but short duration (e.g., common cold), tend to have a weaker influence on prevalence.

    - Longer duration or chronic conditions

        Longer duration or chronic conditions tend to yield higher prevalences. For example, ungueal mycosis, which slowly evolves to cure after treatment start. Similarly, conditions with a longer duration, but with a low incidence (e.g., cystic fibrosis), tend to have a stronger influence on prevalence.

- Migratory movements

    Individuals can either move from the geographic location where the disease prevails or toward it, therefore canceling or starting their participation in prevalence counting, respectively. Depending on their status—case or population—, they might be counted as numerators or denominators of formulas (2.1) and (2.2), respectively, thereby influencing prevalence.

Choosing between incidence or prevalence to measure disease frequency can sometimes be a matter of feasibility or convenience. Incidence deals with new cases of a disease, and in some instances, it is difficult to determine exactly when a given ailment has started, specially a chronic one. For example, glaucoma starts as an asymptomatic disease, often diagnosed during a routine ophtalmologic consultation. If an investigator wished to measure glaucoma incidence, it would be necessary to determine when exactly did a particular case started and that would be clinically impossible. An intraocular pressure screening program would be an option, but a complicating endeavor. Measuring prevalence—simply counting glaucoma cases regardless if they are new or not—could be a surrogate for incidence in this case.

CHAPTER 3

Health indicators

Health indicators are summary measures meant to detail specific epidemiological aspects of human health, as well as health systems performance.

3.1 Preamble

In most instances, health indicator studies involve dynamic populations. Based on actuarial data, it is assumed that, if the corresponding deaths and/or drop-outs are meant to happen, they do so at half the folow-up time. Similarly to disease frequency measures (Chapter 2: Disease Frequency Measures), it is generally recommended: (1) using multiples of 10 such as 100, 1000, 10,000, etc. as denominators in health indicators formulas and (2) assuming years as the time unit. This normatization allows: (1) direct comparisons between different populations, (2) contextualization according to the size of the geographic area under study (e.g., 100,000 for a city or country, 1000 for an outbreak), and (3) adjustments according to survival or mortality due to some rare condition (e.g., using a denominator of 1,000,000 for a rare disease). Alternatively, mortality rates can be expressed with person-time units as denominator instead of population size, for example, 544 deaths/100,000 persons-year.

3.2 Survival

Survival expresses the probability of an individual NOT dying due to a disease, within a timespan $[(t_0, t)]$. This health indicator is actually an extensiton of cumulative incidence index (Chapter 2: Disease Frequency Measures), since it measures the probability of an individual who belongs to a population surviving to a given illness. Survival is determined by the formula:

$$S_{(t_0,t)} = 1 - CI_{(t_0,t)} \tag{3.1}$$

where $S_{(t_0,t)}$ is survival within a timespan, and $CI_{(t_0,t)}$ is cumulative incidence within a timespan.

Practical Biostatistics
DOI: https://doi.org/10.1016/B978-0-323-90102-4.00001-1

For example, cumulative incidence for respiratory infection in male smokers older than 65 years old diagnosed during a 7 days interval admission to XYZ Hospital ICU is 0.043. By applying formula (3.1):

$$S_{(t_0,t)} = 1 - 0.043 = 0.957$$

The above result means that the probability of an older than 65 years male smoker admitted at XYZ Hospital ICU surviving to a respiratory infection contracted during a 7 days interval admission is 95.7%.

3.3 Mortality

In collective health setting, mortality is a term which has no precise definition. It is used to describe the occurrence of death in human populations. Mortality indicators, on their turn, are usually calculated according to the following basic parameters: (1) a numerator that expresses the number of deaths, (2) a denominator that expresses the size of the studied population, and (3) an implicit timespan $[(t_0, t)]$. In dynamic populations, the interpretation of mortality indicators can be influenced by the number of births, newcomers, or by individuals who dropped out from the considered population. Take the following epidemiologic data from ABC City as example:

- mortality indicator (January 1972): 544 deaths/100,000 people-year
- mortality indicator (January 1973): 544 deaths/100,000 people-year
- population growth rate (January 1972 to January 1973): 10%

Presuming ABC City population increased during this time interval, one can assume that ABC City mortality indicator has actually decreased instead of stabilizing, even though 1972 and 1973 mortality indicators are the same.

3.3.1 Crude mortality rate

Crude mortality rate is the simplest form of mortality indicator. It expresses the plain number of deaths along a timespan in a particular population, sized at the middle of this timespan. It is determined by the formula:

$$CMR_{(t_0,t)} = D_x / P_S \qquad (3.2)$$

where $CMR_{(t_0,t)}$ is crude mortality rate, D_x is number of deaths along the timespan, and P_S is population sized at the middle of the timespan.

For example, 384 deaths occurred in ABC City from January to August of 2002, in a population sized at the middle of this timespan as 61,000 people. By applying the formula:

$$CMR_{(t_0,t)} = \frac{384}{61,000} = 0.006$$

In plain terms, the above result means that an individual at ABC City had a chance of 6:1000 (or 0.6%) of dying from January to August of 2002. Nothwithstanding, this information can only be applied to a specific circumstance, that is, ABC City general population from January to August of 2002. It would not be useful for comparison purposes regarding data from different contexts. Nevertheless, crude mortality rate can be standardized in order to allow comparisons, through general mortality coefficient. It is determined by the formula:

$$GMC_{(t_0,t)} = \frac{D_x}{P_S} \times 1000* \div \left(\frac{x**}{12}\right) \tag{3.3}$$

where $GMC_{(t_0,t)}$ is general mortality coefficient, D_x is number of deaths along the timespan, P_S is population sized at the middle of the timespan, * is a multiple of 10 (in this example, 1000), and **x is period of time. The choice regarding the time unit must be contextualized, for example, *year* for a city or country or *month* for an outbreak (for the above formula, chosen time unit is *year*, and it was adapted accordingly, i.e., as monthly fractions).

Results can be expressed, for example: as number of deaths/per 1000 people-year. Using the above case as example:

$$GMR_{(t_0,t)} = \frac{884}{61,000} \times 1000 \div \left(\frac{8}{12}\right) = 24.1$$

3.3.2 Specific mortality rate

Specific mortality rate expresses mortality rate relatively to a specific event or population type. It is determined by the formula:

$$SMR_{(t_0,t)} = D_x / SP_S \tag{3.4}$$

where $SMR_{(t_0,t)}$ is a specific mortality rate, D_x is a number of deaths along the timespan, and SP_S is a specific event or population type.

Examples of specific mortality rates are:
- maternal mortality rate: 82 maternal deaths/10,000 deliveries
- under 5 mortality rate: 12 deaths/1000 live births
- ER admission mortality rate: 28 deaths/1000 ER admissions

Specific mortality coefficients can also be determined, just as for general mortality coefficient (item 3.1).

3.3.3 Mortality proportion

Mortality proportion expresses the proportion of deaths in a given population and is determined by the formula:

$$MP_{(t_0,t)} = D_x / N' \tag{3.5}$$

where $MP_{(t_0,t)}$ is mortality proportion, D_x is number of deaths, and N' is number of individuals at the beginning of the timespan.

For example: 384 deaths occurred in ABC City from January to August of 2002 and its population in January 1st was 61,000 people. By applying the formula:

$$MP_{(t_0,t)} = \frac{384}{61,000} = 0.006 \ (\text{or} \ 0.6\%)$$

3.3.4 Lethality rate

Lethality rate expresses the death rate related to a specific condition among individuals who bear it. It is usually noted as percentage and is determined by the formula:

$$LR_{(t_0,t)} = D_x/ND \tag{3.6}$$

where $LR_{(t_0,t)}$ is lethality rate, D_x is number of deaths, and ND is number of diseased individuals.

For example, lethality rate related to meningococcemia in malnourished infants in a subsaharian country: 0.023 (or 2.3%).

3.3.5 Mortality indicators according to cause of death

Mortality indicators according to cause of death describe death epidemiology associated with a specific cause or group of causes. Two types of mortality indicators according to cause of death will be detailed here.

3.3.5.1 Mortality proportion due to cause of death

Mortality proportion due to cause of death expresses the number of deaths due to a specific cause in relation to the total number of deaths due to all causes, occurring in a given geographic area. It is determined by the formula.

$$MPCD = ND_{s(t_0,t)}/ND_{ac(t_0,t)} \tag{3.7}$$

where $MPCD$ is mortality proportion due to cause of death, $ND_{s(t_0,t)}$ is number of deaths due to a specific cause, and $ND_{ac(t_0,t)}$ is number of deaths due to all causes.

For example, 51 deaths due to hemorrhagic stroke occurred in a total of 892 deaths due to all causes in a year at ABC City. By applying the formula:

$$MPCD = \frac{51}{892} = 0.057 \ (\text{or} \ 5.7\%)$$

3.3.5.2 Mortality rate due to cause of death

Mortality rate due to cause of death provides an estimate of death risk in the general population from a given geographic area. It is determined by the formula.

$$MRCD = ND_{s(t_0,t)}/P_S \times 100,000* \tag{3.8}$$

where *MRCD* is mortality rate due to cause of death, $ND_{s(t_0,t)}$ is number of deaths due to a specific cause, P_S is population sized at the middle of the timespan, and * is a multiple of 10 (in this example, 100,000).

For example, 51 deaths due to hemorrhagic stroke occurred in a population of 650,000 people in a year at ABC City. By applying the formula:

$$MRCD = \frac{51}{650,000} \times 100,000 = 7.84$$

(or 7.84 deaths due to hemorrhagic stroke per 100,000 people)

3.4 Life indicators

3.4.1 Life expectancy

Life expectancy correponds to the quantity of life in years someone is expected to live. It is used to measure the ability of a society in prolonging human life. There are three outcome categories generally used in a follow-up survival study: deceased, alive, or censored (lost to follow-up, switched treatment, dropout, etc.). Nevertheless, alive and censored are considered as one, for the purpose of final analysis. Life expectancy can be expressed through the so-called life table, whose analysis demands three assumptions: (1) no seasonal variation regarding mortality risk, (2) withdrawals are independent of mortality risk, and (3) mortality risk remains constant within the study intervals (yearly, in general).

Life expectancy can be described according to the following parameters.

3.4.1.1 Simple survival

Simple survival is the most elementary parameter regarding life expectancy and the least accurate. It can by its turn be expressed by three different indicators: (1) mean survival (total of years lived/number of patients), (2) median survival, and (3) overall survival rate. Taking life table depicted in Table 3.1, one can exemplify:
- mean survival: total of years lived (45.4 years)/number of patients (18) = 2.5 years
- median survival = 2.1 years
- overall survival rate: 9 surviving patients among 18 patients = 50%

3.4.1.2 Person-year survival rate and person-year death rate

Person-year survival rate is determined by the formula:

$$PYSR = 1 - PYDR \tag{3.9}$$

where *PYSR* is person-year survival rate, and *PYDR* is person-year death rate.

Taking Table 3.1 study as example:

$$PYSR = 1 - 0.132 = 0.868$$

Table 3.1 Life table built on survival data regarding a lung cancer type, spanning 6 years.

Patient	2002	2003	2004	2005	2006	2007	Outcome	Years*
1							alive	3.9
2							alive	5.6
3							withdrawn	2.3
4							death	4.2
5							alive	5.2
6							death	3.2
7							death	1.9
8							death	3.9
9							alive	3.3
10							withdrawn	1.8
11							alive	2.9
12							death	1.0
13							withdrawn	1.2
14							death	1.6
15							alive	0.8
16							alive	1.2
17							alive	0.8
18							alive	0.6
PERSON-YEARS								45.4

*Years of life, measured along the duration of the study.

Notes: Notice that patients are also recruited along this timespan and not only from its start as it would be the case in an analytical study (observational or intervention studies).

Person-year death rate, on its turn, is determined by the formula:

$$PYDR = number\ of\ followed\ years/person\text{-}years \qquad (3.10)$$

where *PYDR* is person-year death rate.

Taking Table 3.1 study as example:

$$PYDR = \frac{6}{45.4} = 0.132$$

3.4.2 Years of potential life lost

Years of potential life lost (YPLL) estimates lifetime, according to the years of life that could have potentially been lived by individuals in a population, had death not occurred before life expectancy age (here arbitrated as 75 years). This indicator is the simple summation of the lost years of life of individuals belonging to a certain population (Table 3.2). YPLL complements life expectancy parameter, since the former expresses the number of "stolen" years from the lifetime of individuals of the considered population.

Table 3.2 Lost years of life of a population of 10 individuals (YPLL = 253).

Individuals	Age of death	Life expectancy	Lost years of life
1	22	75	$(75 - 22) = 53$
2	63		12
3	74		1
4	65		10
5	12		63
6	34		41
7	65		19
8	51		24
9	45		30
10	75		0
YPLL	—	—	**253**

It is also possible to calculate YPLL per death. This indicator expresses the average number of years that could have been lived by each individual of the studied population and can be calculated according to the formula:

$$YPLL \ per \ death = life \ expectancy - average \ age \ at \ death \qquad (3.11)$$

Taking Table 3.2 as example:

$$YPPL = 75 - 50.6 = 24.4$$

YPLL is expected to be directly proportional to life expectancy (i.e., the higher life expectancy is, the higher the number of "stolen" years is expected to be).

3.5 Morbidity indicators

Incidence and prevalence are morbidity measures, but detailed elsewhere in this book.

CHAPTER 4

Epidemiological studies

Epidemiological studies aim to establish the frequency of a condition in a given population. In this type of study, establishing reliable cause—effect correlations stands beyond investigators' possibilities, assuming precise correlation between environmental exposure or nonexposure, and the observed phenomenon can not be established. For this reason, epidemiological studies remain mostly limited to measuring frequencies. Possible cause—effect correlations should be clarified through analytical studies.

Epidemiological studies can be classified in three types.

4.1 Ecological studies

Ecological studies are populational database surveys meant to highlight possible correlations between an environmental factor and an observed condition in a defined geographic area. For example, researchers conclude that during the last 3 years, students from school A located in an underprivileged district have been presenting a higher frequency of upper respiratory virus infection, comparatively to students from school B located in a middle class district, whose frequency is considered usual. Ecological studies goals can be classified as follows:

- general goal: to assess how social and environmental elements affect collective health, through the combination of different databases of great populations
- specific goals: (1) to generate etiologic environmental hypotheses regarding a specific condition, and (2) to measure the efficacy of given interventions on a population (e.g., vaccination)

4.1.1 Variable types

The types of studied variables in ecological studies can be classified as following:

- Aggregated measures

    Aggregated measures summarize characteristics of the studied population, under a common basis. For example, proportion of women with polycystic ovarian disease, incidence or intestinal parasitism, proportion of heavy drinkers, etc.

- Environmental measures

    Environmental measures detail specific aspects of the environment where the studied population is inserted. For example, water bacterial concentration in a

Practical Biostatistics
DOI: https://doi.org/10.1016/B978-0-323-90102-4.00010-2

public school, radioactiviy rates in a postwar zone, chemical contamination rates in residential buildings, etc.
- Global measures
    For example, demographic density and total population.

4.1.2 Methods

According to study design, ecological studies can be performed through the following methods.

4.1.2.1 Exposure measurement methods

Exposure measurement methods are oriented for exploring specific aspects of the exposure itself. They can be subclassified into (1) analytical method (investigated exposure is methodically measured, e.g., immunophenotyping a virus strain from a flu outburst), and (2) exploratory method (exposure is found serendipituously, i.e., along the investigation process).

4.1.2.2 Clustering methods

Clustering methods are oriented for expanding the knowledge basis on the exposure regarding time and space dimensions. They can be subclassified into:
- Temporal series studies
    Temporal series studies goal is to identify temporal patterns potentially associated to an epidemiologic factor of environmental significance, by clustering temporal data either from different populations or from a single population. For example, could two smallpox epidemics occurred 2 months apart in the same State be related?
- Multiple group studies
    Multiple group studies goal is to identify spacial patterns potentially associated to an epidemiologic factor of environmental significance, by clustering space data either from different populations or from a single population. For example, could two smallpox epidemics occurred in two different cities 100 miles apart be related?
- Mixed studies
    Mixed studies combine the characteristics and goals of the former clustering methods study subtypes.
    Ecological studies can also be used to measure the efficacy of population interventions. Their advantages are: (1) low cost, (2) speed of performance, and (3) detection of subtle differences among regions, due to the their broad geographic coverage. Their limitations are: (1) impossibility of controlling for confounders, (2) variation of information quality due to source diversity, (3) long latency diseases, and (4) heterogeneous environmental factor and disease distribution among different populations. Due to these limitations, the environmental effect on the disease can not be estimated.

Nevertheless, it is still possible to perform regression studies (Chapter 19: Correlation and Regression).

4.2 Cross-sectional studies

In cross-sectional studies, the frequency of a given condition in a naturally evolving population under a suspected exposure factor is analyzed, like a snapshot, through a cross-section (Fig. 4.1).

Advantages of cross-sectional studies are: (1) suitability for studying the frequency of long duration conditions with an undetermined onset in a greater population, and (2) in principle, it can be started anytime. Limitations of cross-sectional studies are: (1) it is not possible to know precisely the onset of an investigated condition in a particular individual of the target population, therefore bias is inevitable (for that reason, cross-sectional studies are not amenable for quantifying risks and odds, like in observational studies), (2) considered exposure factors are generally permanent characteristics, such as sex gender or ethnicity, and (3) survival selective bias (only individuals with the condition who survived will be studied). Cross-sectional studies are generally performed through questionnaires.

Hybrid studies models can also be performed:
- Cohort prevalence study

Cohort prevalence studies are useful whenever a given condition either is not or can not be associated with a given exposure factor in the beginning of a cohort study, but nevertheless becomes associated with it later on study progression. At this point, a cross-sectional analysis can be applied in order to rescue the already collected data. For example, in a cohort study, investigators aim to establish the

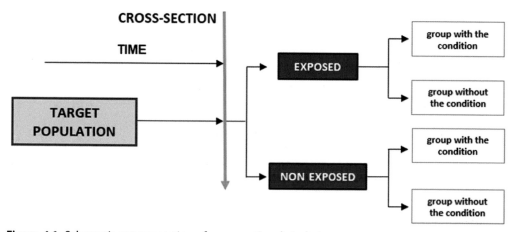

Figure 4.1 Schematic representation of cross-sectional study type.

risk of bladder cancer with smoking, but at a given point, an unexpected exposure factor is identified.

- Selected prevalence study

     Selected prevalence studies can be used whenever there is a cohort to be followed and specific individuals from this cohort are selected for cross-sectioning. For example, women with supraventricular arrhythmia are selected from an original cohort of women with gestational diabetes to be cross-sectioned at the end of the study.

- Sectional follow-up study

     Sectional follow-up studies can be used whenever the investigators of an ended cross-sectional study desire to follow-up the individuals that have not contracted the studied condition, but still remain exposed. For example, bone marrow transplant patients who have not acquired graft-versus-host disease at the end of the cross-sectional study, but are followed thereafter in order to estimate the incidence of the condition.

- Repeat panel study

     Repeat panel studies are useful to assess the effects of an exposure factor over a dynamic population. Assuming altering the composition of a dynamic population demands time, the cross-section must be performed repeatedly at given intervals on the so-called panel, that is, the sample of the dynamic population studied at that particular interval. This type of hybrid study is suitable for investigating the effects of broad exposure factors such as vaccination or environmental pollutants.

4.3 Longitudinal studies

In longitudinal studies, a cohort is followed for several years, sometimes decades, in order to establish the frequencies of specific conditions and their correlation with environmental or other biological factors. Comparisons can be performed either in-between subjects or between different subjects. A classic example is the Framingham Heart Study, which began in 1948 with 5209 subjects and is currently in its third generation of participants. Knowledge on important environmental factors, currently associated to cardiovascular risk such as lifestyle (smoking, diet, and exercise) and aspirin use, is derived from this study.

CHAPTER 5

Pharmacoeconomics

Pharmacoeconomics is a discipline that endeavors to optimize the outcome of a health intervention for the expenses invested. It allows objective assessment of the following parameters, altogether: (1) the cost of a therapeutic intervention, (2) efficacy and safety outcomes, (3) patient preferences and the quality of life provided, (4) labor and indirect economic repercussions attributed to the intervention, and (5) financial return provided by the health investment, for the patient as well as society. Due to the human aspects involving this matter, the interpretation of pharmacoeconomic data should be performed on a subjective—and not objective—basis.

5.1 Costs and benefits

Costs and benefits can be categorized as follows:

5.1.1 Personal

- Costs: (1) direct (medicines, medical consultations, lab workout, procedures in general), (2) indirect [transportation, general support (aged, handicapped)], and (3) imponderable (human distress associated with the condition);
- Benefits: (1) direct (patient health and well-being), (2) indirect (productivity gain due to the intervention), and (3) imponderable (relieved human distress due to the intervention).

5.1.2 Institutional

Institutional costs can be categorized into healthcare direct costs (e.g., hospital admissions) and nonhealthcare costs (e.g., housing, prison system). Regarding the former category, the following accounting resources can be used for cost estimation:

- daily cost: it expresses the mean daily cost per admitted patient. It is the most accurate of all institutional accounting resources;
- condition-specific daily cost: it expresses the mean daily cost per admitted patient diagnosed with a specific pathological condition;
- diagnosis-related group daily cost: it expresses the mean daily cost associated with a group of pathological conditions (e.g., heart failure syndromes);
- microcost: it expresses the cost of each individual item used in a patient (e.g., hypodermic needles, oxygen catheters, tablets);

Practical Biostatistics
DOI: https://doi.org/10.1016/B978-0-323-90102-4.00019-9

- cost:reimbursement ratio: it expresses the relation between what was spent by the institution and what was reimbursed. It is always supposed to be ≤ 1.

5.2 Cost-oriented timing

Timing considerations oriented for cost minimization can be assessed through cost standardization and discount.

5.2.1 Costs standardization

Costs standardization helps to estimate costs for the current year, based on the costs practiced in previous years. Feasible methods are (1) determining the quantity of units (exams, syringes, catheterisms, etc.) used in the late year and multiplying it by the cost per unit prevalent in the current year and (2) multiplying the total cost of the late year by the inflation of the current year.

5.2.2 Discount

Discount refers to the avoided cost for condition management, attributable to an intervention, on a yearly basis. It is determined by the formula:

$$\text{discount} = (\text{non-discounted cost}) - (\text{discounted cost}) \tag{5.1}$$

where non-discounted cost = hypothetical yearly future cost WITHOUT the intervention; discounted cost = hypothetical yearly future cost WITH the intervention.

Non-discounted cost can be estimated according to the current yearly cost. Discounted cost, on its turn, is determined by the formula:

$$\text{discounted cost} = \text{annual current cost}/(1+r)^t \tag{5.2}$$

where r represents estimated yearly discount rate (generally 5% or 0.05); ^{t} represents quantity of future years.

r represents the spared value attributable to the intervention. ^{t}, on its turn, does not merely represent the quantity of future years: (1) it has implicit yearly interest rates as if the spared value attributable to the intervention had been invested in another activity instead of condition management, such as the stock market and (2) it increases as years go by because it is expected that the spared value tends to increase with years of intervention management. For example, we wish to determine the discount attributable to a given intervention for the 2018−21 period. Its current yearly cost is U$ 5000. Both non-discounted cost and discounted cost must then be calculated:

$$\text{non-discounted cost:} (2018/\text{U\$ } 5000) + (2019/\text{U\$ } 5000) + (2020/\text{U\$ } 5000)$$

$$+ (2021/\text{U\$ } 5000) = \text{U\$ } 20,000$$

Discounted cost is then calculated for each year:
- 2018:

$$U\$\,5000/(1+0.0)^1 = U\$\,5000$$

- 2019:

$$U\$\,5000/(1+0.05)^1 = U\$\,4700$$

- 2020:

$$U\$\,5000/(1+0.05)^2 = U\$\,4500$$

- 2021:

$$U\$\,5000/(1+0.05)^3 = U\$\,4300$$

$$\text{discounted cost} = U\$\,5000 + U\$\,4700 + U\$\,4500 + U\$\,4300 = U\$\,18,500$$

By applying the discount formula:

$$U\$\,20,000 - U\$\,18,500 = U\$\,1500$$

The above means that adopting the intervention could represent savings of U\$ 1500 for managing the condition.

5.3 Costs minimization analysis

Cost minimization analysis allows a simple cost comparison between or among similar resources that provide similar efficacy and safety outcomes. Once the investigator assumes that the benefits provided by such interventions are equivalent, the patient is expected to prefer the lowest-cost resource.

5.4 Cost–efficacy analysis

Cost–efficacy analysis measures the cost of an achieved benefit by some natural unit (e.g., cost for one less mmHg of diastolic blood pressure, cost for symptom improvement per patient). It is the most commonly found type of pharmacoeconomic resource in the literature, for being quite understandable for healthcare professionals and managers. Cost–efficacy analysis can be performed through the following resources.

5.4.1 Cost—consequence analysis

Cost—consequence analysis corresponds to the simple presentation of an obtained benefit for a given cost. For example, U$ 109 for an opioid used for 1 month in exchange for 9 days without pain.

5.4.2 Cost—efficacy ratio

Cost—efficacy ratio expresses the benefit provided by an intervention, relative to its cost. It can be detailed in three ways:

5.4.2.1 Simple cost—efficacy ratio

Simple cost—efficacy ratio expresses the relationship between cost—consequence analysis parameters. For example, U$ 109/9 days without pain = *U$ 12.10 daily without pain*.

5.4.2.2 Cost—efficacy ratio for percentual unit of success

Cost—efficacy ratio for percentual unit of success expresses possible cost increment due to a given intervention. Examples:

Intervention A: (1) total cost of U$ 55,500 for treating 100 patients (U$ 550 for individual patient, assuming a hypothetical 100% success rate) and (2) 90% actual success rate → U$ 5500 was "wasted" in 10 patients. We ask: how much this "wasting" would have influenced cost, considering the actual success rate?

$$\text{CER for percentual unit of success} = \text{intervention A indivudual cost}$$

$$/\text{intervention A actual success rate*} \quad (5.3)$$

where CER represents cost—efficacy ratio; * represents 'expressed as decimals'.

$$\text{CER for percentual unit of success} = \frac{U\$\ 500}{0.90} = U\$\ 610$$

The above means that each individual patient successfully treated with intervention A would imply a U$ 60 cost increment (U$ 610−U$ 550).

Intervention B: (1) total cost of U$ 40,000 for treating 100 patients (U$ 400 for individual patient, assuming a hypothetical 100% success rate) and (2) 60% actual success rate → U$ 16,000 was "wasted" in 40 patients. We ask: how much this "wasting" would have influenced cost, considering the actual success rate?

$$\text{CER for percentual unit of success} = \text{intervention B individual cost}$$

$$/\text{intervention B actual success rate*} \quad (5.4)$$

where CER represents cost—efficacy ratio; * represents 'expressed as decimals'.

$$\text{CER for percentual unit of success} = \frac{\text{U\$ } 400}{0.60} = \text{U\$ } 660$$

Conclusion: each individual patient successfully treated with intervention B would imply a U\$ 260 cost increment (U\$ 660—U\$ 400).

5.4.2.3 Cost—efficacy ratio for percentual of an additional success

Cost—efficacy ratio for percentual of an additional success expresses the additional total cost of an intervention relative to the additional number of patients who were benefited by that intervention. It is determined by the formula.

$$\text{CER for percentual of an additional success} = \frac{(\text{intervention A individual cost}) - (\text{intervention B individual cost})}{(\text{intervention A actual success rate*}) - (\text{intervention B actual successrate*})} \quad (5.5)$$

where CER represents cost—efficacy ratio; * represents 'expressed as decimals'.

Using intervention A and intervention B cases from the preceding Subchapter as example:

$$\text{CER for percentual of an additional success} = \frac{\text{U\$ } 550 - \text{U\$ } 400}{0.90 - 0.60} = \text{U\$ } 500$$

Since intervention A benefited 90 patients and intervention B benefited 60 patients, one can assume the former benefited 30 additional patients relative to the latter. So, if intervention A costed U\$ 550 for each patient it benefited, then U\$ 550 × 30 patients = U\$ 16,500. Therefore intervention A has costed U\$ 16,500 more than intervention B for benefiting 30 additional patients.

5.4.3 Cost—efficacy increment ratio

Cost—efficacy increment ratio expresses the relativeness between cost and efficacy of two intervention modalities. It can be expressed in two settings.

5.4.3.1 As simple difference between endpoints

Cost—efficacy increment ratio as simple difference between endpoints is used whenever both cost and benefit can be aligned, that is, both are either advantageous or disadvantageous. An example is depicted in Table 5.1.

Intervention B represents a fewer cost while providing two additional days without epigastric pain → the dominant intervention. On the other hand, intervention A represents a greater cost while providing less days without epigastric pain → the dominated intervention.

5.4.3.2 By applying cost—efficacy increment ratio itself

Cost—efficacy increment ratio is used whenever cost and benefit differences are uneven, that is, one is advantageous, whereas the other is disadvantageous. It is determined by the formula.

$$CEIR = \frac{(\text{intervention A cost}) - (\text{intervention B cost})}{(\text{benefit provided by intervention A}) - (\text{benefit provided by intervention B})}$$

$$(5.6)$$

where CEIR indicates cost—efficacy increment ratio.

An example is depicted in Table 5.2.

Intervention A provides 7 days without epigastric pain, nevertheless, it is more costly than intervention B. On the other hand, intervention B is cheaper but provides 2 fewer days without epigastric pain. Assuming we aim to benefit our patients with 2 more days without epigastric pain, no matter the cost, we can determine the latter by applying the formula:

$$CEIR = \frac{U\$\,250 - U\$\,210}{7 \text{ days} - 5 \text{ days}} = U\$\,20 \text{ per additional day without epigastric pain}$$

The above result means that the 2 additional days without epigastric pain will cost U$ 40.

5.4.4 Using graphic resources

Cost—efficacy grid—example in Table 5.3. Cost—efficacy plan—example in Fig. 5.1

Table 5.1 Costs and benefits provided by intervention A × intervention B.

	Intervention A	Intervention B
Cost	U$ 250.00	U$ 210.00
Days without epigastric pain	5 days	7 days

Table 5.2 Costs and benefits provided by intervention A × intervention B.

	Intervention A	Intervention B
Cost	U$ 250.00	U$ 210.00
Days without epigastric pain	7 days	5 days

Table 5.3 A hypothetical cost–efficacy grid, suggesting conducts regarding cost and efficacy.

Cost–efficacy	Inferior cost	Same cost	Superior cost
Inferior efficacy	Apply CEIR		Dominated intervention
Same efficacy		Arbitrary	
Superior efficacy	Dominant intervention		Apply CEIR

CEIR, Cost–efficacy increment ratio.

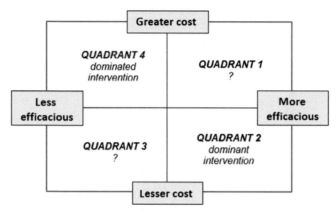

Figure 5.1 A hypothetical cost–efficacy plan.

5.5 Utility

Utility can be defined as the quality of life provided by the intervention, under the context of the condition of interest. It is assessed by the patient according to variables whose nature is familiar to him or her (e.g., degree of pain, restrictions to daily living activities, time for improvement). As so, subjectiveness is its main limitation. Utility is measurable through utility analysis and cost–utility analysis.

5.5.1 Utility analysis

Utility analysis extends function-based quality of life instruments applicability (e.g., Medical Outcomes Short-Form 36, Quality of Well-Being Scale) by including psychologic, socioeconomic, and labor issues, as well as other topics that might influence the quality of life of an individual. This parameter can be assessed through three different methods: (1) scaling, (2) standardized game, and (3) time negotiation.

5.5.1.1 Scaling

Different conditions are scaled from the least to the most severe state (perfect health equals 1.0, whereas dead equals 0.0). For example, allergic rhinitis 0.9, ankle spraining 0.6, and myocardial infarction 0.1. Scaling advantages are (1) many different conditions can be detailed for the same individual, (2) can be performed with no contact with the patient, and (3) less difficult to perform than standardized game and time negotiation (further in the text). Scaling limitations are (1) disregards time and (2) people tend not to group conditions in the extremes of the scale.

5.5.1.2 Standardized game

Standardized game is based on the principle that the patient might either be willing to trade time for better health or remain in the same health status but to live more. This is about a game between the physician and the patient, in which the goal is to reach the so-called *indifference point* (Fig. 5.2).

Patient must choose between keeping the current status or submitting to procedure, as follows:

- the patient is informed that he/she has a 10% chance of total cure with the procedure and 90% chance of dying → patient prefers the procedure;
- the patient is informed that he/she has a 20% chance of total cure with the procedure and 80% chance of dying → patient still prefers the procedure;
- the patient is informed that he/she has a 30% chance of total cure with the procedure and 70% chance of dying → patient hesitates → *indifference point* is reached.

$$\text{utility scoring} = \text{indiference point} = 70\% \ (0.7)$$

Limitations of the method are (1) it might be difficult to perform by the patients, (2) it is applicable only for chronic conditions, and (3) few of these conditions could evolve to "total cure."

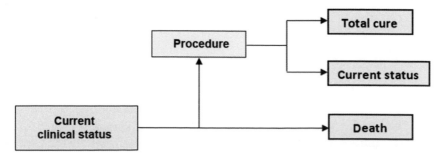

Figure 5.2 Basic scheme for the standardized game between physician and patient.

5.5.1.3 *Time negotiation*

Similarly to standardized game, time negotiation is based on the principle that the patient might either be willing to trade time for better health or remain in the same health status but to live more. This is about a negotiation between the physician and the patient, in which the goal is scoring utility, as follows: the patient must choose between either sticking to the intervention, but with a permanent sequel during t time, or without this permanent sequel during x time $<t$:

- the patient is informed that he/she will either live for 50 years (t) with a permanent sequel or 25 years (x) without this permanent sequel → patient prefers t time;
- the patient is informed that he/she will either live for 50 years (t) with a permanent sequel or 35 years (x) without this permanent sequel → patient still prefers t time;
- the patient is informed that he/she will either live for 50 years (t) with a permanent sequel or 45 years (x) with this permanent sequel → patient hesitates → *indifference point* is reached.

Utility scoring will be determined by the formula.

$$\text{utility scoring} = x/t \tag{5.7}$$

By applying the former example.

$$\text{utility scoring} = \frac{45 \text{ years}}{50 \text{ years}} = 0.9$$

Advantages of the method are (1) it deals with time in an easier fashion than scaling and (2) it is more adaptable than standardized game. Applicability exclusively for chronic conditions is its limitation.

5.5.2 Cost—utility analysis

In cost—utility analysis, health gain provided by an intervention is measured by a nonnatural unit which combines (1) utility, (2) survival (as number of years saved), and (3) cost. The most frequently used nonnatural unit is the QALY (quality-adjusted life-year) score system, where expected life-years are weighted according to their respective quality (best quality possible equals 1, whereas dead equals 0). QALY allows comparisons between different outcomes (e.g., heart disease and perinatal care) for the endpoint is QALY and not the outcome itself. It is determined by the formula.

$$\text{QALY} = \text{saved years of life} \times \text{utility for each saved year of life} \tag{5.8}$$

QALY calculation can be exemplified (Table 5.4).

$$\text{QALY for intervention A} = 5 \times 0.8 = 4.0$$

$$\text{QALY for intervention B} = 7 \times 0.5 = 3.5$$

Table 5.4 An example of QALY calculation.

	Intervention cost	Saved years of life	Utility scoring for each saved year of life
Intervention A	U$ 10,000.00	5	0.8
Intervention B	U$ 20,000.00	7	0.5

Utility gain = (utility for each saved year of life by intervention A) − (utility for each saved year of life by intervention B) (in this example, utility gain = 0.8 − 0.5 = 0.3). (*see Subchapter 5.5.2.1 for utility gain definition*).

QALY assessment can be extended through following two different methods:

5.5.2.1 Utility gain

Utility gain corresponds to the gain in quality of life for each saved year of life provided by an intervention, relatively to another intervention. It is determined by the formula.

$$\text{utility gain} = (\text{utility for each saved year of intervention A})$$
$$- (\text{utility for each saved year of life of intervention B}) \qquad (5.9)$$

By applying the former example:

$$\text{utility gain} = 0.8 - 0.5 = 0.3$$

5.5.2.2 Cost−utility ratio

Cost−utility ratio integrates obtained QALY with the intervention cost in dollars, therefore expressing how much was spent per QALY gained. It is determined by the formula.

$$\text{CUR} = \text{cost of the intervention}/\text{QALY} \qquad (5.10)$$

where CUR represents cost−utility ratio
Cost−utility ratio calculation can be exemplified, based on Table 5.4.

$$\text{intervention A ACU} = \frac{\text{U\$ 10,000}}{\text{4 QALY}} = \text{U\$ 2500/QALY}$$

$$\text{intervention B ACU} = \frac{\text{U\$ 20,000}}{\text{3.5 QALY}} = \text{U\$ 5700/QALY}$$

We infer that intervention A provides more quality of life per saved year of life for a lower cost than intervention B. It should be noted, however, that only when the

QALY of the two interventions is the same, should cost prevail regarding the decision of which intervention is to be adopted.

Cost-effectiveness analysis measures the resources expended in exchange for a given health benefit (e.g., days without pain, vision improvement, muscle gain). Some authors consider cost-effectiveness analysis and cost—utility analysis as synonymous. In cost-effectiveness analysis, natural units and traditional clinical trial endpoints are used.

5.6 Financial resources

Financial resources are applied whenever economic emphasis is needed. Once their parameters are financially expressed, intervention as well as outcomes of different natures can be compared. Their limitation is that cultural acceptance of attributing a monetary perspective to human health is sometimes low.

5.6.1 Cost—benefit analysis

Cost—benefit analysis provides monetary significance for a health intervention by focusing on invested value (cost) in exchange for an attained financial return (benefit). Some cost—benefit analysis measurements are as follows:

- liquid benefit and liquid cost

$$\text{liquid benefit} = \text{total benefits} - \text{total costs} \tag{5.11}$$

$$\text{liquid cost} = \text{total costs} - \text{total benefits} \tag{5.12}$$

Interventions are considered as having a positive cost—benefit relation if liquid benefit >0 and/or liquid cost <0.

- cost—benefit and benefit—cost ratios

$$\text{cost-benefit ratio} = \text{total costs}/\text{total benefits} \tag{5.13}$$

$$\text{benefit-cost ratio} = \text{total benefits}/\text{total costs} \tag{5.14}$$

Interventions are considered as having a positive cost—benefit relation if cost-benefit ratio <1 and/or benefit—cost ratio >1.

5.6.2 Human capital

Human capital expresses indirect and intangible costs related to a condition as well as indirect and intangible benefits related to the respective intervention. It is calculated according to the value of a labor day, either lost to the former or rescued by the latter. For the purposes of human capital assessment, we can consider four types of work:

(1) lost work, (2) lost domestic work, (3) restricted work (compromised productivity), and (4) astray work (next of kin who had to leave his/her work to caregive the patient). Advantages and limitations of human capital are, respectively:

- advantages: (1) straightforwardness, (2) informations can be retrieved from public database, and (3) the values of the four types of work can be estimated with relative facility;
- limitations: (1) biased for certain population types (e.g., unemployed people), (2) it presumes that people who do not work do not provide economic gain (e.g., children, the aged), and (3) it presumes that the daily cost of a health benefit is equivalent to the daily gain provided by some work.

5.6.3 Return rate

Return rate corresponds to return over investment of a health intervention, as compared to a conventional financial investment used as reference.

5.7 Health-related life quality—using questionnaires

Assuming the ultimate goal of an intervention is to improve a patient's quality of life, one can propose that HRQoL (health-related quality of life) assessments can be potentially useful for pharmacoeconomic decision-making routines. HRQoL can be estimated through self-measurement instruments, presented as questionnaires. These are usually elaborated for the two main human health domains, that is, physical and mental health. The following parameters can be researched:

- Generic

    Generic questionnaires are applicable to conditions of different nature. They can be short or long and include (1) physical performance (daily living activities, movement restriction, pain), (2) psychological functionality, (3) social and labor functionality, and (4) general health perceptions.
- Condition-specific

    As the term implies, condition-specific questionnaires are addressed to a certain condition of interest. A cardiology HRQoL questionnaire, for example, could include (1) the presence or absence of palpitations, (2) a family history consistent with cardiovascular risk, and (3) the amount of salt consumption.

5.7.1 Assessing an HRQoL questionnaire

Ideally, an HRQoL questionnaire should be assessed to test its robustness. The following parameters can be used.

5.7.1.1 Reproductibility

Reproducibility tests if HRQoL questionnaire has the capacity of providing consistent answers. Some useful tests are as follows:

- Test—retest consistency

    Test—retest consistency is used to assess similarities between scores for the same health condition, recorded at different times.

- Internal consistency

    Internal consistency cross-checks coherence among answers for different questions, meant for similar endpoints (e.g., vitality vs work readiness).

- Interviewers' consistency

    Interviewers' consistency verifies coherence among answers provided by different individuals, who nevertheless are under the same research setting (e.g., the patient and his/her caregiver).

5.7.1.2 Validity

Validity rectifies HRQoL questionnaire capacity in reflecting reality. It can be assessed through the following tests:

- Content validity

    Content validity assesses if efficacy endpoints in an HRQoL questionnaire properly represent the studied condition (a different questionnaire can be used as reference).

- External criteria validity

    External criteria validity assesses if HRQoL questionnaire is coherent with selected external parameters. For example, a population that scored high in an HRQoL questionnaire is expected to present a low mortality rate.

- Concept validity

    Concept validity rectifies an HRQoL questionnaire efficacy by cross-testing it with different HRQoL questionnaires. It consists of a set of validation tools, as follows:

    - Convergence validation

        Convergence validation is used to assess the results from different HRQoL questionnaire types, regarding the same health domain. For example, results from an HRQoL questionnaire on general mental health must be consistent with the results from another HRQoL questionnaire on a specific mental condition, for the same population.

    - Discriminatory validation

        An HRQoL questionnaire meant for a certain health domain is NOT expected to consistently reproduce the results of another HRQoL questionnaire meant for a health domain of a different nature. So, discriminatory validation is used to assess if an HRQoL questionnaire meant for a certain health domain does NOT consistently reproduce the results of a HRQoL questionnaire meant for a health domain of a different nature. For example, if an HRQoL questionnaire on the physical health domain does NOT consistently reproduce the results of an HRQoL questionnaire on mental health, then the former can discriminate between both domains.

- Known groups validation
  Known groups of different natures are expected to yield different results, even when under the same domain. For example, results related to the prelabor anxiety domain between a group of nulliparous women and a group of nonnulliparous women are expected to be different. Known group validation assures HRQoL questionnaire is sufficiently robust to discriminate between groups, regarding this aspect.

- Responsivity
  Responsivity tests the capacity of a HRQoL questionnaire in showing health status changes, as expressed through differences among evolving scores and among patients with the same conditions but at different severity levels.

5.7.2 Pharmacoeconomics and health-related quality of life questionnaires

The results themselves of a HRQoL questionnaireare not in general sufficiently accurate to be used as parameters for pharmacoeconomic decisions. Nevertheless, they may positively influence the latter, under the following conditions: (1) HRQoL questionnaire favors intervention A against intervention B, (2) intervention A cost ≤ intervention B cost, and (3) intervention A is noninferior to intervention B, regarding efficacy and safety. Under different conditions, only cost—consequence analysis data (*Subchapter 5.4.1*) are expected to be detailed, leaving possible interpretations at the investigator's discretion.

5.8 Decision analysis

The purpose of decision analysis is to determine the likely final cost of two different interventions, whenever working with undetermined parameters is necessary: (1) success or failure, (2) occurrence of an unpredicted adverse reaction, and (3) its cost. Based on this analysis, a decision is made on which would be the most advantageous intervention. Decision analysis steps are as follows:

STEP 1: detailing the interventions

For example, we have two possible interventions, as expressed in Table 5.5. Which intervention would be better, based on the likely final cost of each one?

STEP 2: drawing analysis tree

Table 5.5 Intervention A and intervention B on a comparison basis (percentages expressed as decimals).

	Intervention A	Intervention B
Basic cost (per patient)	U$ 700.00	U$ 500.00
Success probability	0.90	0.80
Failure probability	0.10	0.20
Probability of adverse reaction nonoccurrence	0.90	0.85
Probability of adverse reaction occurrence	0.10	0.15
Cost estimate for treating an adverse reaction (per patient)	U$ 1000.00	U$ 1000.00

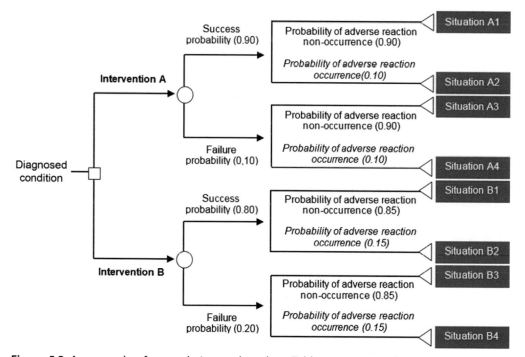

Figure 5.3 An example of an analysis tree, based on Table 5.5 content. The square represents a *decision node*—either intervention A or intervention B, the circle represents the *chance node* for each intervention, whereas the triangle represents the *terminal node*.

An example of an analysis tree is illustrated in Fig. 5.3.
STEP 3: performing calculations

$$\text{situation A1: } 0.90 \times 0.90 = 0.81 \times \text{U\$ } 700 = \text{U\$ } 567 \text{ per patient}$$

$$\text{situation A2: } 0.90 \times 0.10 = 0.09 \times (\text{U\$ } 700 + \text{U\$ } 1000) = \text{U\$ } 135 \text{ per patient}$$

situation A3: $0.10 \times 0.90 = 0.09 \times$ U\$ 700 = U\$ 63 per patient

situation A4: $0.10 \times 0.10 = 0.01 \times$ (U\$ 700 + U\$ 1000) = U\$ 17 per patient

LIKELY FINAL COST (A1 + A2 + A3 + A4) = U\$ 782 per patient

situation B1: $0.80 \times 0.85 = 0.68 \times$ U\$ 500 = U\$ 340 per patient

situation B2: $0.80 \times 0.15 = 0.12 \times$ (U\$ 00 + U\$ 1000) = U\$ 180 per patient

situation B3: $0.20 \times 0.85 = 0.17 \times$ U\$ 500 = U\$ 85 per patient

situation B4: $0.20 \times 0.15 = 0.03 \times$ (U\$ 500 + U\$ 1000) = U\$ 45 per patient

LIKELY FINAL COST (B1 + B2 + B3 + B4) = U\$ 650 per patient

Conclusion: intervention B would probably cost less than intervention A. Notwithstanding, intervention B would be less efficacious than intervention A (0.80 success probability vs 0.90 success probability, respectively) → perform CEIR (Subchapter 4.3.2):

$$\frac{\text{U\$ } 782 - \text{U\$ } 650}{0.90 - 0.80} = \text{U\$ } 1320 \text{ per additional success with intervention A}$$

STEP 4: performing sensitivity analysis

In decision analysis context, sensitivity analysis will be useful for detecting the point where both interventions would yield equal costs, according to either the probability of adverse reaction occurrence or the cost estimate for treating an adverse reaction. Both are manipulated within a simulation range, until a threshold where probable final costs of both interventions match.

Bibliography

Suggested reading (Part 2)

Bodrogi, J., Kaló, Z., 2010. Principles of pharmacoeconomics and their impact on strategic imperatives of pharmaceutical research and development. Br. J. Pharmacol. 159 (7), 1367–1373.

Brown, G.C., Brown, M.M., 2016. Value-based medicine and pharmacoeconomics. Dev. Ophthalmol. 55, 381–390.

Canadian Medical Association Journal, 2017. Basic statistics for clinicians. www.cmaj.ca. (Accessed 4 May 2017).

CDC/World Fund Programme, 2005. A Manual: Measuring and Interpreting Malnutrition and Mortality. UNHRC.

Centre for Evidence Based Medicine, 2016. Statistics for the Clinic. Department of Medicine, Toronto General Hospital. www.cebm.utoronto.ca (Accessed 3 May 2016).

Estrela, C., 2001. Metodologia Científica—Ensino e Pesquisa em Odontologia, first ed. Editora Artes Médicas, Porto Alegre (Chapter 6).

Everitt, B., 2006. Medical Statistics From A to Z. A Guide for Clinicians and Medical Students, second ed. Cambridge University Press, London.

Everitt, B.S., et al., 2005. Encyclopaedic Companion to Medical Statistics. Hodder Arnold, London.

http://sphweb.bumc.bu.edu/otlt/MPH-Modules/EP/EP713_DiseaseFrequency/EP713_DiseaseFrequenc4.html. Disease frequency measures. (Accessed 3 May 2016).

Hulley S.B., et al., 2001. Designing Clinical Research: An Epidemiological Approach, second ed. Williams Wilkins, Philadelphia, PA.

Medronho, R.A., et al., 2009. Epidemiologia, second ed. Atheneu, Rio de Janeiro.

Rascati, K.L., 2009. Essentials of Pharmacoeconomics. Wolters Kluwer Health, Philadelphia, PA.

Sackett, D.L., et al., 2001. Evidence-Based Medicine: How to Practice and Teach EBM, second ed. Elsevier Health Sciences, Amsterdam.

uwphi.pophealth.wisc.edu/publications/issue-briefs/issueBriefv05n07.pdf (Accessed 20 January 2017).

Wesley, D., 2007. Life table analysis. J. Insur. Med. 30 (4), 247–254.

www.cdc.gov/ophss/csels/dsepd/ss1978/lesson3/section2.html (Accessed 28 April 2016).

PART 3

Observational studies

The objective of Part III is to expand our knowledge basis on observational studies and to introduce derivative concepts: odds ratio, relative risk, and number needed to harm. Models for increasing the accuracy of this study type are also described.

CHAPTER 6

Basic concepts in observational studies

The main objective of observational studies is to establish the degree of hazard for a certain condition in relation to a considered exposure factor. In these types of study the frequency of a condition in a population is observed under so-called natural conditions. As such, active intervention from the investigator on its evolution is not applicable. Submitting the observed population to "real world" situations is their advantage. Their limitation is the yielding of less accurate conclusions, assuming uncontrolled variables and potential confounders can generate bias. Observational studies can be classified as case-control and cohort studies.

6.1 Case-control studies

In case-control studies, two groups are retrospectively compared, according to the following model: (1) one group WITH the condition (case) is subdivided into two subgroups—one exposed and the other nonexposed to a studied exposure factor, and (2) another group WITHOUT the condition (control) is subdivided into two subgroups—one exposed and the other nonexposed to the same factor (Fig. 6.1).

Case-control studies aim to determine the *odds* in acquiring a condition, under exposure to a considered factor. For example, miners have 1.5:1 odds in presenting asbestos-associated lung fibrosis, relatively to the general population. Their advantages are a better financial affordability in comparison to cohort studies (Section 6.2) and feasibility for immediate performance, since they are generally retrospective. Their limitation is the poor control over the exposure factor, uncontrolled variables, and potential confounders, for the latter reason. Given the fact they focus on the outcome and "move backward" to the exposure factor, they are generally retrospective. Their inferred association strength—odds ratio—is determined by a specific formula (Chapter 7: Determination of Association Strength Between an Exposure Factor and an Event in Observational Studies).

6.2 Cohort studies

In cohort studies, a cohort of healthy subjects is divided into two groups, according to exposure or nonexposure to a given factor—exposed cohort and nonexposed cohort -, in principle for prospective follow-up. At the end of the study, the number of subjects with and without the condition is measured for both (Fig. 6.2).

Practical Biostatistics
DOI: https://doi.org/10.1016/B978-0-323-90102-4.00022-9

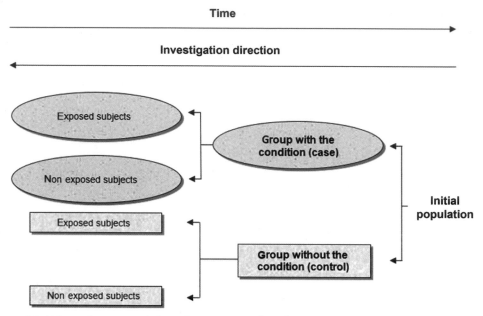

Figure 6.1 Schematic representation of a case-control study type.

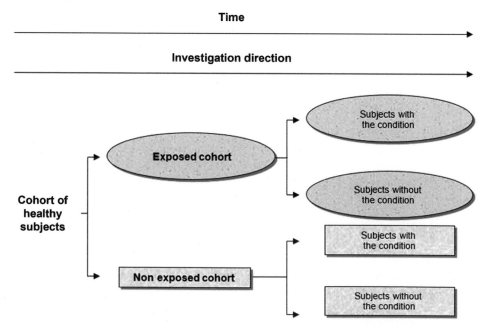

Figure 6.2 Schematic representation of a cohort study type.

Cohort studies aim to determine the *risk* of acquiring a condition under exposure to a considered factor (exposure can either happen simultaneously with study start or might have already happened before it). For example, nuclear power plant workers have a 2.5 greater risk of presenting high-grade lymphoma relative to a nonexposed cohort. The advantage of this type of study is affording a better control over exposure level, covariates, and potential confounders, since they are prospective. Their limitations are as follows: (1) need for waiting so that exposure factors exert their effects, (2) follow-up loss, and (3) higher cost. Given the fact they focus on the exposure factor and "move forward" to the outcome, they are generally prospective.

Nonexposed cohort can be selected from three source types, depending on the nature of the studied object and resources available:

* Internal

    Originated from the same setting as the exposed cohort. It is the preferred source type whenever the studied condition is a very specific one and/or a high degree of similarity between exposed and nonexposed cohorts is highly desirable. For example, a cohort of female employees routinely exposed to radiation therapy versus a nonexposed cohort of females employees working in the same building, both hired in the same year and belonging to the same age range, for measuring breast cancer relative risk (see further).

* External

    Originated from a different setting than the exposed cohort. Preferred source type whenever the studied condition is more broadly prevalent and/or a high degree of similarity between exposed and nonexposed cohorts is not absolutely necessary. For example, a cohort of aged people routinely exposed to sunlight versus a nonexposed cohort which lives indoors belonging to a different region, for measuring relative risk for skin melanoma in face and forearms.

* General population

    Originated from the general population. Preferred source type whenever the studied condition is broadly prevalent and/or organizing a nonexposed cohort is difficult. For example, a cohort of measles vaccinated children versus a nonexposed cohort that lives in a different country (not vaccinating in a country where vaccination is mandatory would not be feasible), for measuring relative risk of contracting measles.

Their inferred association strength—relative risk—is determined by a specific formula (Chapter 7: Determination of Association Strength Between an Exposure Factor and an Event in Observational Studies). Note: the expression "cohort study" refers to a specific type of observational study; in this sense, the term "cohort" must be differentiated from its broader sense (see Glossary).

In retrospective cohort studies, two groups are retrospectively identified and "prospectively" compared, according to the following model: a cohort of healthy subjects

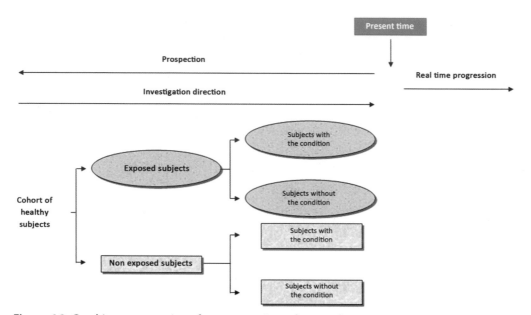

Figure 6.3 Graphic representation of a retrospective cohort study type.

is subdivided into two groups: (1) one exposed to a given factor, and (2) the other nonexposed to the same factor (Fig. 6.3).

The advantages of retrospective cohort studies include a better affordability compared to cohort studies and feasibility for immediate performance, since they are retrospective. Their limitation is the poor control over the exposure factor, covariates, and potential confounders, for the latter reason.

CHAPTER 7

Determination of association strength between an exposure factor and an event in observational studies

The goal in observational studies is to measure the odds or the risk for the occurrence of an event, between two groups. According to the observational study type, different approaches are possible.

7.1 Case-control studies

7.1.1 Odds ratio

The odds ratio (OR) is an index for association strength determination between an exposure factor and an event. In a general context the term represents the ratio of probabilities of the two possible states of a binary variable in one group relative to another (e.g., probability of symptomatic remission in group A, against probability of symptomatic worsening in group B). In observational studies setting, OR expresses the ratio between the odds for the occurrence of an event in a group exposed to a factor and the odds for the occurrence of the same event in a group exposed to a different factor (or not exposed). OR can be used in studies of epidemiological interest or in therapeutic observational studies. OR can also derive the number needed to harm (NNH) (*Subchapter 7.1.2*).

7.1.1.1 *For studies of epidemiological interest*

For example, a population of 100 individuals is divided into a group with lung cancer (case) and a group without lung cancer (control), with the aim of measuring the *odds* for the occurrence of lung cancer related to smoking exposure. Both groups are subdivided into two subgroups each—smokers and nonsmokers—and retrospectively followed for up to 25 years (Fig. 7.1 and Table 7.1).

Based on the above results, we can infer:

- There was a 4:1 odds of smokers presenting lung cancer (a/b).
- There was a 1:4 odds of nonsmokers presenting lung cancer (c/d).
  Establishing the OR, according to the formula:

$$OR = \frac{(a/b)}{(c/d)} \tag{7.1}$$

Practical Biostatistics
DOI: https://doi.org/10.1016/B978-0-323-90102-4.00024-2

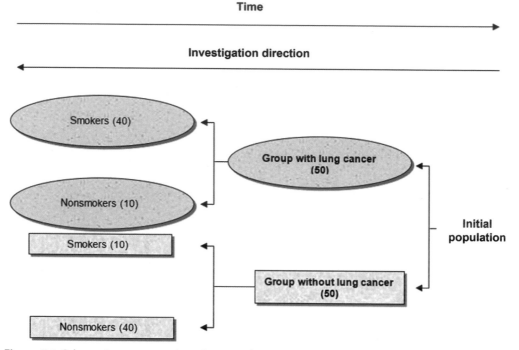

Figure 7.1 Schematic representation of an initial population of healthy individuals and lung cancer patients, under a case-control study type.

Table 7.1 Study results for odds calculation.

| | Results | |
	Case	Control
Smokers	40 (*a*)	10 (*b*)
Nonsmokers	10 (*c*)	40 (*d*)

$$OR = \frac{40/10}{10/40} = 16$$

This result means that the *odds* of lung cancer occurrence was 16:1 for smokers in relation to nonsmokers.

7.1.1.2 For therapeutic studies

For example, a population of 120 perimenopausal women is divided into a group with perimenopausal symptoms (case) and a group without perimenopausal symptoms (control), with the aim of measuring the *odds* for the occurrence of perimenopausal symptoms relatively to regular ingestion of soy isoflavones. The groups are subdivided into two subgroups each: (1) subgroup A—women who regularly ingest soy isoflavones—and (2) subgroup B—women who do not ingest soy isoflavones. Both groups are retrospectively followed for up to 10 years (Fig. 7.2 and Table 7.2).

Based on the above results, we can infer:

- There was a 0.6:1 odds of women who regularly ingest soy isoflavones in presenting perimenopausal symptoms (*a/b*).
- There was a 2.5:1 odds of women who do not ingest soy isoflavones in presenting perimenopausal symptoms (*c/d*).

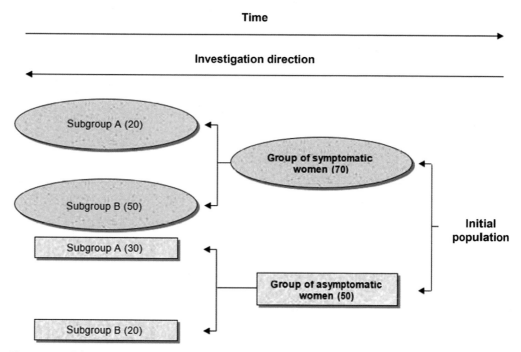

Figure 7.2 Schematic representation of a population of perimenopausal women, under a case-control study type.

Table 7.2 Study results for odds ratio calculation.

| | Results | |
	Case	Control
Subgroup A	20 (*a*)	30 (*b*)
Subgroup B	50 (*c*)	20 (*d*)

Establishing the OR, according to formula (7.1):

$$OR = \frac{20/30}{50/20} = 0.26$$

This result means that the *odds* of perimenopausal symptoms occurrence was 0.2:1 for women who regularly ingest soy isoflavones, in relation to women who do not ingest soy isoflavones.

Case-control studies do not allow *risk*, but *odds* status correlations only, due to the following reasons:

- Assessment orientation

  Case-control studies allow group assessment rather than individual against individual exposure evaluation, assuming they afford less control over study conditions.
- Number of assessed individuals

  Even though cases and controls derive from the same population, their number is arbitrarily defined by the investigator and therefore does not necessarily represent this population. For the same reason, it is not usual to sum cases and controls in case-control study tables.

The above reasons also explain why OR has a limited application in intervention studies (*Part 4: Biostatistics of Intervention Studies - The Clinical Trials*).

Determination of OR cutoff values that point to a significant relationship between the exposure factor and the event, is empirically based. Normally, the following factors are taken into consideration:

- Influence of unknown variables and potential confounders

  Case-control studies are more prone to the occurrence of unknown variables and potential confounders than cohort studies. Therefore there is a greater "tolerance" for higher OR cutoff values comparatively to relative risk (RR) results.
- Event severity

  The more severe the event, the lesser the "tolerance" for the assumption of more elevated OR cutoff values.
- Selection bias

  The very fact cases are affected by the studied condition suggests the former ones might have been more exposed than the controls, generating exposure bias.
- Case-respondent and/or investigator bias

  Possible if the case-respondent is aware of the studied condition, for the individual is expected to have a better memory for possible exposure factors associated with his or her condition. On the other hand, the investigator might influence him- or herself regarding exposure intensity over the case, depending on his or her degree of belief on the putative role of the exposure factor.

Pairing cases and controls can minimize the above limitations. It consists in selecting controls according to certain coincident and potentially confounding characteristics with the cases, for example, gender, demographics, and age. For example, in a coronary heart disease study that aims determining the odds for acute myocardial infarction, smoking cases and controls and nonsmoking cases and controls would be preferentially paired. It is understood that once pairing is adopted, the number of paired cases and controls is supposed to be the same. Pairing limitations are as follows: (1) OR analysis could be performed only between matching pairs, (2) the coincident characteristic becomes an exposure factor itself, and (3) recruiting can take longer.

7.1.2 Number needed to harm

NNH corresponds to the number of individuals that must be treated, so that one individual presents an adverse reaction accountable to the treatment. The main usefulness of NNH is to make the OR data sound more practical to physicians and comprehensible for patients. Its interpretation must be performed based on the physician's own practice experience and NNHs established for other treatment modalities related to the case. For example, a population of 180 individuals with recently treated lung tuberculosis is divided into a group with drug-induced hepatitis (case) and a group without drug-induced hepatitis (control), with the aim of measuring the *odds* for the occurrence of isoniazid related hepatitis. The groups are subdivided into two subgroups: (1) regimen A—treated with isoniazid—and (2) regimen B—not treated with isoniazid. Both groups are retrospectively followed until the beginning of tuberculostatic regimens (Fig. 7.3 and Table 7.3).

Establishing the OR, according to formula (7.1):

$$OR = \frac{65/35}{35/45} = 2.4$$

NNH is determined by the formula:

$$NNH = 1 - [PEER \times (1 - OR)]/(1 - PEER) \times PEER \times (1 - OR) \qquad (7.2)$$

where NNH is the number needed to harm, PEER the patient expected event rate, and OR the odds ratio.

In our example, chosen PEER corresponds to the proportion of regimen B individuals belonging to the drug-induced hepatitis group: 30% (or 0.3).

$$\frac{1 - [0.3 \times (1 - 2.4)]}{(1 - 0.3) \times 0.3 \times (1 - 2.4)} = 5$$

This result means that it would have been necessary to treat five patients with lung tuberculosis, so that one of them presented isoniazid drug—induced hepatitis.

Alternatively, it is possible to consult an NNH table (Tables 7.4 and 7.5).

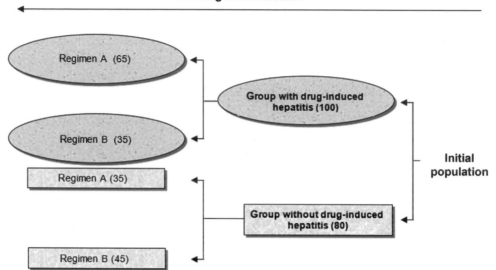

Figure 7.3 Schematic representation of a population of recently treated lung tuberculosis patients, under a case-control study type.

Table 7.3 Study results for odds ratio calculation.

	Results	
	Case	**Control**
Regimen A	65 (*a*)	35 (*b*)
Regimen B	35 (*c*)	45 (*d*)

7.2 Cohort studies

RR is an index for association strength determination between an exposure factor and an event. It is defined as the ratio between the risk for the occurrence of an event in a group exposed to a factor and the risk for the occurrence of the same event in a group exposed to a different factor (or not exposed). RR can be used in studies of epidemiological interest or in therapeutic observational studies. By analogy with OR, RR can also derive the NNH (*Subchapter 7.2.2*).

Table 7.4 Number needed to harm table for odds ratio (OR) <1.0 and >1.0.

		For OR <1.0						
		0.9	0.8	0.7	0.6	0.5	0.4	0.3
PEER	0.05	209	104	69	52	41	34	29
	0.10	110	54	36	27	21	18	15
	0.20	61	30	20	14	11	10	8
	0.30	46	22	14	10	8	7	5
	0.40	40	19	12	9	7	6	4
	0.50	83	18	11	8	6	5	4
	0.70	44	10	13	9	6	5	4
	0.90	101	46	27	18	12	9	4

PEER, Patient expected event rate.

Table 7.5 Number needed to harm table for odds ratio (OR) <1.0 and >1.0.

		For OR >1.0						
		1.1	1.25	1.5	1.75	2	2.25	2.5
	0.05	212	86	44	30	28	18	16
	0.10	113	46	24	16	13	10	9
	0.20	64	27	14	10	8	7	6
PEER	0.30	50	21	11	8	7	6	5
	0.40	44	19	10	8	6	6	5
	0.50	42	18	10	8	6	6	5
	0.70	51	23	13	10	9	8	7
	0.90	121	55	33	25	22	19	18

PEER, Patient expected event rate.

7.2.1 Relative risk

7.2.1.1 For studies of epidemiological interest

For example, a population of 100 individuals is divided into an exposed group (smokers) and a nonexposed group (nonsmokers), with the aim of measuring the *risk* for the occurrence of lung cancer related to smoking exposure. Both groups are prospectively followed for 15 years and then subdivided into two subgroups each—individuals with lung cancer and individuals without lung cancer (Fig. 7.4 and Table 7.6).

Based on the above results, we can infer:

- There is an 80% risk of smoking exposed individuals in presenting lung cancer [$a/(a + b)$].

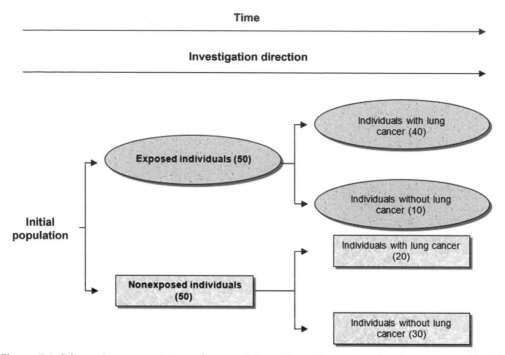

Figure 7.4 Schematic representation of a population of smoking exposed and nonexposed individuals, under a cohort study type.

Table 7.6 Study results for risk calculation.

	Results		
	With lung cancer	**Without lung cancer**	**Total**
Exposed	40 (*a*)	10 (*b*)	50
Nonexposed	20 (*c*)	30 (*d*)	50

- There is a 40% risk of nonsmoking exposed individuals in presenting lung cancer [$c/(c + d)$].

  Establishing the relation between the risks (RR), according to the formula:

$$RR = \frac{[a/(a + b)]}{[c/(c + d)]} \tag{7.3}$$

$$RR = \frac{[40/(40 + 10)]}{[20/(20 + 30)]} = 2$$

This result means that the *risk* of lung cancer is twice as high among smoking exposed individuals in relation to nonexposed individuals.

7.2.1.2 For therapeutic studies

For example, a population of 120 perimenopausal women is divided into group A (women who regularly ingest soy isoflavones) and group B (women who do not ingest soy isoflavones), with the aim of measuring the *risk* for the occurrence of perimenopausal symptoms relatively to regular ingestion of soy isoflavones. Both groups are prospectively followed for 3 years and then subdivided into two subgroups each—symptomatic women and asymptomatic women (Fig. 7.5 and Table 7.7).

Based on the above results, we can infer:
- There is a 28% risk of women who regularly ingest soy isoflavones in presenting perimenopausal symptoms [$a/(a + b)$].
- There is a 60% risk of women who do not ingest soy isoflavones in presenting perimenopausal symptoms [$c/(c + d)$].

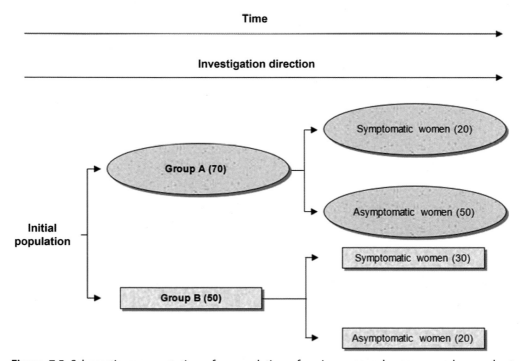

Figure 7.5 Schematic representation of a population of perimenopausal women, under a cohort study type.

Table 7.7 Study results for relative risk calculation.

	Results		
	Symptomatic	Asymptomatic	Total
Group A	20 (*a*)	50 (*b*)	70
Group B	30 (*c*)	20 (*d*)	50

Establishing the relation between the risks, according to formula (7.3):

$$RR = \frac{[20/(20 + 50)]}{[30/(30 + 20)]} \sim 0.5$$

This result means that the *risk* of perimenopausal symptoms occurrence is 0.5 for women who regularly ingest soy isoflavones in relation to women who do not ingest soy isoflavones.

7.2.2 Number needed to harm

NNH corresponds to the number of individuals that must be treated, so one of them presents an adverse reaction accountable to the treatment. The main usefulness of NNH is to make RR data sound more practical to physicians and comprehensible for patients. Its interpretation must be performed based on physician's own practice experience and on NNHs established for other treatment modalities related to the case. For example, a population of 180 individuals with the diagnosis of lung tuberculosis is divided into two groups treated with different regimens, with the aim of measuring the *risk* for the occurrence of isoniazid drug—induced hepatitis: (1) regimen A—isoniazid included—and (2) regimen B—isoniazid not included. Both groups are prospectively followed for 1 year and then divided into two subgroups each: individuals with drug-induced hepatitis and individuals without drug-induced hepatitis (Fig. 7.6 and Table 7.8).

NNH is determined by the formula:

$$\frac{a}{(a + b)} - \frac{c}{(c + d)} \tag{7.4}$$

$$\frac{65}{(65 + 35)} - \frac{35}{(35 + 45)} = 0.22 \rightarrow 22$$

This result means that it would be necessary to treat 22 patients with lung tuberculosis, so that one of them would present isoniazid drug—induced hepatitis.

Time

Investigation direction

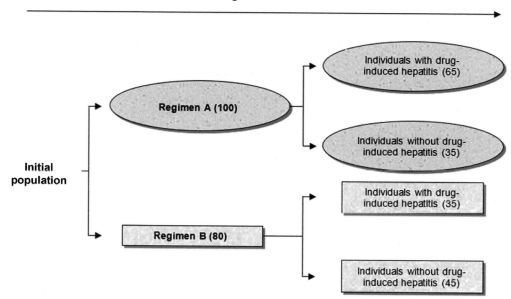

Figure 7.6 Schematic representation of a population of lung tuberculosis patients, under a cohort study type.

Table 7.8 Study results for number needed to harm calculation.

	Results		
	With hepatitis	**Without hepatitis**	**Total**
Regimen A	65 (*a*)	35 (*b*)	100
Regimen B	35 (*c*)	45 (*d*)	80

Cohort studies allow individual to individual rather than group exposure assessment, assuming they afford more control over study conditions. For this reason, this type of study allows *risk* status correlations. An intervention study (*Chapter 9: The Intervention Studies*) can also derive a cohort study, regarding safety aspects of the former (often the endpoint for which RR is the most useful).

Determination of RR values that point to a significant relationship between the exposure factor and the event is empirically based. Generally, three factors are taken into consideration:

• Influence of unknown variables and potential confounders

Cohort studies are less prone to unknown variables and potential confounders' occurrence than case-control studies. Therefore a smaller tolerance for greater RR values is expected.

*   Event severity

    The more severe the event, the lesser the tolerance for greater RR values.

*   Selection bias

    The very fact cases are affected by the studied condition suggests they might be more exposed than the controls, generating exposure bias.

CHAPTER 8

Increasing accuracy in observational studies

An important limitation of observational studies is the poor control exerted by the investigator over study conditions, due to the effects of uncontrolled variables different from the major variable—condition or exposure. These uncontrolled variables either influence (prospective studies) or will have influenced (retrospective studies) the evolution of the study, making interpretation of results more difficult. Notwithstanding, it is possible to minimize this effect by including these uncontrolled variables on study protocol as covariates. Stratified analysis and multivariable analysis are two types of statistical resources, fitted for this purpose:

8.1 Stratified analysis

In stratified analysis, covariates different from the major variables—condition or exposure—are weighted in odds ratio (OR) and relative risk (RR) calculations, respectively. Through stratification of case/control and exposure/nonexposure groups, new data can be inferred [only bivariate analysis (i.e., one covariate at a time against the major variable) will be detailed]. Let us take two different examples, from the former chapter (Chapter 7):

Example 8.1: A population of 120 perimenopausal women presented an OR of 0.26:1 for the occurrence of perimenopausal symptoms for women who regularly ingest soy isoflavones, in relation to women who do not ingest soy isoflavones. General study results are detailed in Table 8.1.

Establishing the OR, according to formula (7.1):

$$OR = \frac{20/30}{50/20} = 0.26$$

Then, both subgroups are stratified according to age ranges:
- 40—50 years old (Table 8.2)

$$OR = \frac{3/10}{20/7} = 0.07$$

Practical Biostatistics
DOI: https://doi.org/10.1016/B978-0-323-90102-4.00004-7

Table 8.1 General study results for odds ratio calculation.

	Results	
	Case	Control
Subgroup A	20	30
Subgroup B	50	20

Table 8.2 Study results according to 40–50 years old age range.

	Results	
	Case	Control
Subgroup A	3	16
Subgroup B	5	2

Table 8.3 Study results according to 51–65 years old age range.

	Results	
	Case	Control
Subgroup A	7	8
Subgroup B	15	3

Table 8.4 Study results according to 66–80 years old age range.

	Results	
	Case	Control
Subgroup A	10	6
Subgroup B	30	15

- 51–65 years old (Table 8.3)

$$OR = \frac{10/8}{10/6} = 0.17$$

- 66–80 years old (Table 8.4)

$$OR = \frac{7/12}{20/7} = 0.8$$

As can be clearly noticed, OR increases across the three proposed age ranges. Had we not performed this stratification, this important information would have remained undetected. Nevertheless, the hypothesis if these results should represent significant differences must be tested through statistical tests (Chapter 14: Hypothesis Testing).

Example 8.2: A population of 100 individuals presented a RR of 2 for the occurrence of lung cancer related to smoking exposure, for smokers in relation to nonsmokers. General study results are detailed in Table 8.5.

Establishing the RR, according to formula (7.3):

$$RR = \frac{[40/(40+10)]}{[20/(20+30)]} = 2$$

Then, both groups are stratified according to family history for lung cancer:
• Negative family history (Table 8.6)

$$RR = \frac{[22/(22+6)]}{[12/(12+16)]} = 1.8$$

• Positive family history (Table 8.7)

$$RR = \frac{[18/(18+4)]}{[8/(8+14)]} = 2.2$$

Table 8.5 General study results for RR calculation.

	Results	
	With lung cancer	**Without lung cancer**
Exposed	40	10
Nonexposed	20	30

Table 8.6 Study results according to a positive family history.

	Results	
	With lung cancer	**Without lung cancer**
Exposed	22	6
Nonexposed	12	16

Table 8.7 Study results according to a negative family history.

	Results	
	With lung cancer	Without lung cancer
Exposed	18	4
Nonexposed	8	14

As can be clearly noticed, RR differs according to the family history of lung cancer. Had we not performed this stratification, this important information would have remained undetected. Nevertheless, the hypothesis if these results should represent significant differences must be tested through statistical tests.

8.2 Multivariable analysis

The reader is advised to study Chapter 19, Correlation and Regression, before exploring the following subchapter.

Multivariable analysis represents a more robust tool for inaccuracies minimization in observational studies, than stratified analysis. Through this approach, the effect of different covariates (independent variables) on study outcome (dependent variable) is firstly considered discretely, then simultaneously, for a more realistic result. Let us illustrate this concept through the example of a case-control study.

With the aim of determining the degree of influence from combined covariates (*independent variables*) on blood glucose levels (*dependent variable*) abnormalities during the first three days in an adult postoperative ICU, the records of 804 patients comprehending a 3 years period were retrospectively investigated. Blood glucose levels from 792 ICU nonsurgical patients, measured during their first three days of admission, served as controls. Multiple linear regression was the type of multivariable analysis applied. Cutoffs for abnormal blood glucose levels were <50 and >140 mg/dL.

First, OR were calculated for each covariate individually, regardless of the other covariates (crude OR; calculations not shown). Then, multiple linear regression analysis was performed, when biological influence of study covariates on each other was weighed, for OR recalculation (adjusted OR; calculations not shown) (Table 8.8).

Crude OR results pointed to age >65, orthopedic surgery, diagnosis of postoperative cytokine storm/sepsis, and preoperative risk assessment, as the covariates more strongly associated to abnormal blood glucose levels. After multivariable analysis, results are consistent with neurosurgery, use of insulin, and preoperative risk assessment, as the covariates are more strongly associated to this abnormality.

Table 8.8 Crude odds ratio (OR) and adjusted OR (OR after multivariable analysis) of 7503 observations related to abnormal blood glucose levels in ICU postoperative patients and 6903 observations related to ICU nonsurgical patients.

	Crude OR	Adjusted OR
Age > 65	1.01[a]	1.02
Gender	1.3	1.0
Abdominal surgery	0.9	1.0
Orthopedic surgery	1.0[a]	0.8
Neurosurgery	1.0	1.3[a]
Diagnosis of preoperative infection	1.2	1.3
Diagnosis of postoperative cytokine storm/sepsis	1.3[a]	1.7
Use of adrenaline (y/n)	1.8	1.4
Use of insuline (y/n)	1.9	1.3[a]
Preoperative risk assessment	1.9[a]	2.0[a]

[a]Strength of association between the variable and blood glucose levels was statistically significant ($P < .05$).

Multivariable analysis is robust enough to eventually identify some covariates as confounders, since its methods can comprehend several covariates simultaneously. However, it cannot adjust for unknown variables. Only through randomization it is possible to compensate for their presence, by evenly distributing them between or among groups.

Bibliography

Suggested reading (Part 3)

Canadian Medical Association Journal, 2017. Basic statistics for clinicians. www.cmaj.ca (Accessed 4 May 2017).

Everitt, B., 2006. Medical Statistics From A to Z. A Guide for Clinicians and Medical Students, second ed. Cambridge University Press, London.

Everitt, B.S., et al., 2005. Encyclopaedic Companion to Medical Statistics. Hodder Arnold, London.

Hulley S.B., et al., 2001. Designing Clinical Research: An Epidemiological Approach, second ed., Lippincott Williams & Wilkins, Philadelphia, PA.

Jaeschke, R., Guyatt, G., Shannon, H., et al., 2007. Assessing the effects of treatment: measures of association. Can. Med. Assoc. J 152, 351–357.

Katz, M.H., 2007. Multivariable Analysis. A Practical Guide for Clinicians. Cambridge University Press, London.

Sackett, D.L., et al., 2001. Evidence-Based Medicine: How to Practice and Teach EBM, second ed. Elsevier Health Sciences, Amsterdam.

PART 4

Biostatistics of intervention studies

The clinical trials

The objective of Part IV is to present the essentials of biostatistics applied to intervention studies (randomized clinical trials), in a step-by-step fashion. The reader is expected to study the steps sequentially, for optimal assimilation.

CHAPTER 9

The intervention studies

Intervention studies (sometimes referred to as "randomized trials" or "controlled studies") are prospective cohort studies, generally performed with a reference group ("control group"). Here, the investigator plans and actively intervenes on the factors influencing his/her cohort, minimizing the influence of uncontrolled variables and potential confounders. They are the most commonly used type of study to determine the efficacy of a drug or vaccine. Intervention studies can be classified according to the applied intervention model.

9.1 Reference standard

9.1.1 Noncomparative studies

In noncomparative studies, tested group has no reference standard to be compared to. Results interpretation is difficult, assuming the investigator can never be sure if the findings represent or not an actual influence from the tested drug or vaccine.

9.1.2 Comparative studies (controlled studies)

In comparative studies, tested group has a reference standard group (control) to be compared to, represented either by an active drug or vaccine known to be efficacious (active or positive control) or by placebo (negative control). Results interpretation by the investigator is more consistent, since there is a comparison reference. In most situations, control group subjects are contemporary to tested group subjects. But, in special circumstances, historical (literature) controls can be used.

9.2 Relation between samples and of a sample with itself

9.2.1 Nonpaired (independent) samples studies

In nonpaired sample studies, individuals from a group can be freely matched for comparison with any individuals from the other group. Comparability is provided by study design itself. For example, a sample of elderly patients admitted for cholelithiasis is divided into a group directed for emergency cholecystectomy and another group directed for conservative cholelithiasis management aiming further elective surgery, for mortality and morbidity rates comparison.

Practical Biostatistics
DOI: https://doi.org/10.1016/B978-0-323-90102-4.00012-6

9.2.2 Paired (dependent) samples studies

In paired samples studies, individuals from a group can be matched for comparison exclusively with specific individuals from the other group. Three types of pairing are possible:

- Self-pairing

    An individual is matched with him/herself. For example, a sample of male patients with chronic neuropathic pain due to lumbar disk herniation complicated with radiculoneuropathy is treated with acupuncture during 3 months and with anticonvulsants during the same interval thereafter, for optimal pain management comparison.

- Natural pairing

    Natural pairing is performed between two different, but extremely linked individuals, such as, for example, monozygotic twins.

- Artificial pairing

    Artificial pairing is performed between independent individuals but paired according to a specific studied variable. For example, a sample of elderly patients admitted for cholelithiasis is divided into a group directed for emergency cholecystectomy and another group directed for conservative cholelithiasis management aiming further elective surgery, for comparison of mortality and morbidity rates. Nevertheless, according to study design, individuals from one group can only be paired with individuals from the other group who share the same serum bilirubin level ranges.

9.3 Awareness of tested drug, vaccine, or exam

9.3.1 Open studies

In open studies, both the investigator and study subject are aware of the nature of the tested drug or vaccine. It is applied whenever it is neither possible nor desirable to conceal it from both. For example, comparison study between PUVA (Psoralen + UVA exposure) versus subcutaneous etanercept for skin psoriasis control. Its limitation is bias generation both from the investigator and the study subject.

9.3.2 Single-blinded studies

In single-blinded studies, the investigator—but not study subject—is aware of the tested drug or vaccine. It is applied whenever it is neither possible nor advisable that the investigator is unaware of the situation. For example, comparison study between dobutamine and milrinone efficacy for pulmonary capillary wedge pressure control, during cardiogenic shock. Its limitation is bias generation due to self-suggestion on the part of the investigator.

9.3.3 Double-blind studies

In double-blind studies, neither the investigator nor the study subject is aware of the tested drug or vaccine. It is applied to avoid bias generation due to self-suggestion from both parties. For example, comparison study between placebo and a type 5 phosphodiesterase inhibitor, for erectile dysfunction management.

9.4 Study subject allocation method

9.4.1 Nonrandomized studies

In nonrandomized studies the investigator selects study subjects to be allocated in one of the study groups, according to preestablished criteria. It is applied whenever study design demands selective allocation. For example, influenza vaccine efficacy study between two groups of subjects of different age ranges (40−50 and 60−70 years). It is susceptible to bias due to misinterpretation of data.

9.4.2 Randomized studies

In randomized studies the investigator randomly allocates study subjects into study groups. The advantages are as follows: (1) random error minimization through even distribution of individual characteristics between or among study groups, (2) systematic error minimization through avoidance of study subjects selectivity by the investigator and (3) minimization of the influence of uncontrolled variables, through their equal distribution between or among study groups.

9.5 Follow-up method

9.5.1 Parallel studies

In parallel studies, study groups progress in parallel along the investigation, until its end.

9.5.2 Crossover studies

In crossover studies the groups swap their respective arms at a specific point of investigation evolution. For example, a new ergot alkaloid derivative + acetaminophen combination showed a significant decrease of migraine symptoms in group A subjects comparatively to group B subjects, who took only acetaminophen. After regimen swap, group B subjects showed the same symptom decrease first associated to group A subjects and the latter presented the same symptoms level that group B subjects did in the first phase of the study (Fig. 9.1).

The advantages of crossover studies include the following: (1) they corroborate the findings of the first phase of the study by reproducing them in the second phase,

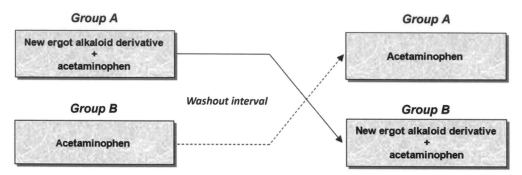

Figure 9.1 Schematic representation of a crossover study.

reinforcing the final conclusion of the study and (2) they afford comparison between groups in a self-paired fashion, that is, groups A and B can be compared against themselves at the end of the study and this allows for a greater biological homogeneity. Crossover studies limitation is the need for a washout interval between study phases.

9.6 Subgroup analysis

Subgroup analysis corresponds to the assessment of a particular subgroup in a clinical trial, often motivated by a serendipitous finding. For example, an investigator notices that a given tested analgesic unexpectedly provides better results in elderly than in younger patients, then deciding to promote a derivative analysis between both subgroups. The limitation of subgroup analysis is its propensity for generating type I errors due to the potential influence of confounders which might have positively influenced the initial finding.

Study design is the combination of the awareness level of tested drug or vaccine, chosen comparative reference, study subject allocation and follow-up methods, study planned duration and groups and subgroups quantity and exams involved, all adjusted to answer the investigator's hypothesis. A randomized, double-blind, crossover and placebo-controlled trial would be an example of a study design.

CHAPTER 10

n Estimation and *n* assessment of a published trial

One of the most sensitive tasks in Biostatistics is to determine *n* or sample size. This is so because, ideally, a clinical trial should comprehend the totality of individuals in the world carrying the specific condition considered, in order to apply its results with the highest possible level of certainty. Nevertheless, carrying such a task would be prevented by impracticality and the risks of incorrectly collecting and handling the massive amount of data that would be generated. Therefore, a convenient resource would be sampling from a population, based on an assumed ideal *n*, with the aim of further extrapolating biostatistical test results back to this population, with a reasonable degree of safety.

Even though there is no "correct" or "incorrect" *n*, sampling process should assure that the sample is minimally representative of the original population. To achieve this goal, sampling should ideally adhere to the following guidelines.

Note: determination of the parameters presented in this chapter can be performed through widely available online resources. We will detail corresponding calculations in a conventional manner, for didactical purposes.

10.1 Factors influencing n determination

There are two types of factors that influence *n* determination, which must be taken into account before proceeding with its estimate.

10.1.1 Empirical factors

Empirical factors correspond to subjective and logistic considerations weighted by the investigator and the biostatistician and are imprecise in nature. The following factors are usually considered:
- historical data from previous studies;
- data from pilot studies, in case there are no historical data;
- data derived from experimental models;
- effect size (Section 10.1.2.2) which might represent a relevant difference, from a clinical standpoint;
- biological characteristics of the studied condition;

Practical Biostatistics
DOI: https://doi.org/10.1016/B978-0-323-90102-4.00013-8

- feasible recruiting rate in the research center(s);
- study type (e.g., crossover studies probably demand a smaller n, for the reasons enlisted in Section 9.5 of Chapter 9: The Intervention Studies);
- eligibility criteria (narrow eligibility criteria lead to a more homogenous population, probably affording a smaller n);
- condition prevalence and frequency of corresponding complications; and
- time available for study completion.

10.1.2 Mathematical factors

Mathematical factors afford a more precise approach to n determination, even though it will still be an estimate.

10.1.2.1 Statistical power of the test

Statistical power of the test corresponds to the capacity of a given statistical test (Chapter 14: Hypothesis Testing) in effectively finding a difference between two compared groups, in other words, the capacity of rejecting H_0 (null hypothesis) (Chapter 1: Investigator's Hypothesis and Expression of Its Corresponding Outcome: Measures for Results Expression of a Clinical Trial) whenever it is false. If it is not sufficiently powerful, truly existing differences might not be detected, making the investigator incur a type II error. Obviously, the smaller the odds a detected difference should represent an error, the greater the odds this difference actually exists. As such, this power could be expressed in a probabilistic format, through the formula:

$$\text{Statistical power of the test} = 1 - \beta \qquad (10.1)$$

β is the statistical significance level corresponding to the highest degree of tolerability for a statistical test in *not* detecting a difference between two compared groups (0.20) (an alternative way of expressing this information is that the tolerance for type II error is 20%). So, statistical power of the test could be calculated as follows:

$$1 - 0.20 = 0.80 \ (\text{or } 80\%)$$

Therefore 80% would be the lowest acceptable degree of probability of a given statistical test in finding a difference between the groups of a study. Elevating n could by itself increase this probability. In summary, the higher n, the higher the probability of a statistical test in finding a difference between groups and by inference its power.

Could we make our test statistically more "powerful" by decreasing β to the same value as α (0.05)? Yes but, if we did that, statistical power of the test would rise up to 0.95 (or 95%), considerably increasing the difficulty in rejecting null hypothesis. This would push *n* to such high figures that it would probably make most clinical trials unviable.

10.1.2.2 Effect size

Effect size corresponds to the difference between the results of two groups. For example, a sample of female patients with hyperthyroidism is divided into two groups, with free serum thyrotoxin levels as the study endpoint, measured at the end of the trial: (1) group A—treated with a test antithyroid drug (2.1 ng/dL) and (2) group B—treated with a reference drug (22.0 ng/dL). Effect size is 19.9 ng/dL.

There are some effect size associated factors that can influence its impact on *n* determination:

• Effect size magnitude

Effect size magnitude corresponds to the magnitude (or size) of yielded difference between the results of two groups. Depending upon the condition studied and the trial goals, an adequate *n* will be necessary to yield a clinically significant effect size between study groups. As a general rule, the smaller the effect size magnitude expected to yield a clinically significant difference, the larger *n* must be and vice versa. If effect size does not correspond to a sufficient magnitude to yield a clinically significant difference and a small *n* is adopted, the likelihood for type II error is expected to be higher.

• Dispersal degree of study results

If yielded results are excessively disperse, it can become difficult to detect differences between results of different groups. For example, troponin T serum levels (ng/mL) from a group of 5 subjects with myocardial infarction treated with thrombolytics 1 hour after symptoms onset (group A), are compared with the results from another group of 5 subjects with the same diagnosis, treated 4 hours after symptoms onset (group B) (Fig. 10.1).

Notice how group A results are more disperse in comparison to group B. Under such circumstances, it would be difficult to assign the detected difference between groups to either the early administration of thrombolytics or to a mere variability of results. Test power would, therefore, be diminished. Increasing *n* could compensate for this limitation, by diluting variability.

• Test direction

Test direction refers to the types of answers an investigator's hypothesis allows. It can be of two types:

• One-sided

Only one type of result direction is possible. For example, could surgical antibiotic prophylaxis decrease post-surgical infection incidence?

• Two-sided

Two opposing answers are possible. For example, what would be the effect of surgical antibiotic prophylaxis on post-surgical infection incidence (would it

Group A

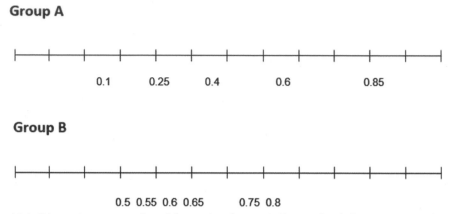

Group B

Figure 10.1 Schematic representation of the results of troponin T serum levels from group A and group B.

either increase *or* decrease it)? Bilateral studies demand a higher *n* to make an effect size attainable.

10.1.2.3 p

The smaller the assumed cutoff for *p* (generally 0.05) is, the higher *n* might be.

10.1.2.4 Refusals and drop-outs

One must always take refusals and drop-outs into account during *n* determination for a trial. The corresponding *n'* can be determined by the formula:

$$n' = n/(1 - q) \tag{10.2}$$

where *n'* is the *n* after taking refusals and drop-outs into account, *q* the expected proportion of refusals and drop-outs. For example—during a study planning, *n* of 65 is estimated and a proportion of 10% of refusals and drop-outs is expected:

$$n' = \frac{65}{(1 - 0.10)} = 72$$

where *n'* should be 72 so that the study finishes with 65 subjects.

10.1.2.5 Sampling error tolerability

The higher the tolerability for error (generally $\pm 5\%$) is, the higher *n* might be.

10.1.2.6 N

As a general rule, the greater *N* (the number of individuals in a population) the least influence on *n* estimation it is expected to have. This is so because one can assume that in a great population, sampled individuals can be replaced by "new" ones as the former ones are selected. On the other hand, this presumption cannot be adopted in a smaller *N*.

10.2 *n* Estimate

Obs.: the following methods are applicable for equal sized groups.

10.2.1 For studies aiming to analyze differences between means

To analyze the differences between means, the following two steps are performed.

10.2.1.1 Standardized difference determination

Comparing two different effect sizes requires considering them in the context of the intrinsic variability of the studied endpoint. For example, a 20 IU/L mean drop of AST (aspartate aminotransferase) serum levels in a context of a standard deviation (Chapter 12: Measures for Results Expression of a Clinical Trial) of ± 30 IU/L, is more significant than a 20 IU/L mean drop in a context of a standard deviation of ± 60 IU/L [a wider standard deviation points to the existence of higher discrete (i.e., individual, single) serum AST measurements].

Standardized difference is an index that allows an estimate of the relevance of this difference. It combines the target difference—the minimal effect size considered as clinically relevant expected to be found in the trial-to-be—and the established standard deviation for the studied endpoint. Obviously, standard deviation could not be known in advance of study performance. Therefore it must be determined based on historical data from similar studies, pilot studies, or on empirical clinical grounds.

Standardized difference is determined by the formula:

$$\text{Standardized difference} = \frac{\text{Target difference}}{\text{Standard deviation}} \qquad (10.3)$$

For example—we wish to determine the standardized difference for a test antihypertensive drug, considering a blood pressure of 15 mmHg as the target difference, in a ± 25 mmHg standard deviation context:

$$\text{Standardized difference} = \frac{15}{25} = 0.6$$

10.2.1.2 n Determination itself

- By applying a sample size nomogram

  Test power, standardized difference, and the chosen significance level are applied to a sample size nomogram. By applying the above example (Fig. 10.2): (1) test power = 80% (or 0.80), (2) standardized difference = 0.6, and (3) $p = .05$.

  n would be approximately 80 (40 individuals to each group).

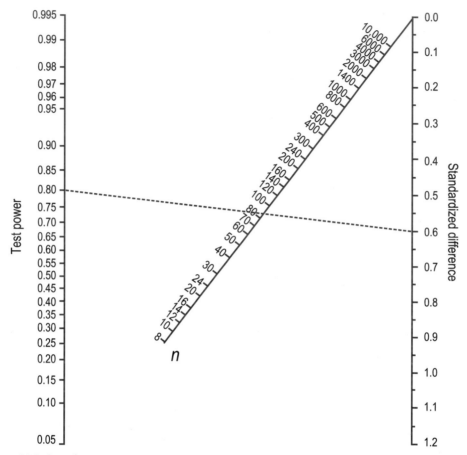

Figure 10.2 Sample size nomogram adjusted for $p < .05$.

- By applying a sample size formula for n estimate of an individual group

$$n_i = (2/d^2) \times C_{p,\text{power}} \qquad (10.4)$$

where n_i is the n of an individual group, d the standardized difference, and $C_{p,\text{power}}$ the constant defined by p and test power. $C_{p,\text{power}}$ can be determined according to Table 10.1.

$$n_i = \frac{2}{0.6^2} \times 7.9 = 44$$

n_i would be 44 (therefore n would be 88).

Table 10.1 $C_{p,power}$ determination according to test power and p.

p	TEST POWER			
	50	80	90	95
0.05	3.8	7.9	10.5	13.0

Note: Sample size nomogram and sample size formula do not necessarily yield exactly same values, assuming their objective is to give *n estimates*.

10.2.2 For studies aiming to analyze differences between proportions

For example, according to historical and empirical data, we infer that the expected 1 year survival rate for pleural mesothelioma treated with a cisplatin plus gemcitabine regimen is 42%, against a 32% rate with an oxaliplatin plus raltitrexed regimen. We wish to estimate *n* to develop a formal study on this issue, comparing group A (cisplatin plus gemcitabine) against group B (oxaliplatin plus raltitrexed). It can be performed in two steps.

10.2.2.1 Standardized difference determination

Standardized difference is determined by the formula:

$$\text{Standardized difference} = \frac{p_A - p_B}{\sqrt{p'(1 - p')}} \tag{10.5}$$

where p_A is the group A proportion, p_B the group B proportion, and p' the arithmetic mean of $(p_A + p_B)$.

$$\text{Standardized difference} = \frac{0.42 - 0.32}{\sqrt{0.37(1 - 0.37)}} = 0.2$$

10.2.2.2 n Determination itself

- By applying a sample size nomogram

    Test power, standardized difference, and the chosen significance level are applied to a sample size nomogram. By applying the above example (Fig. 10.3): (1) test power = 80% (or 0.80), (2) standardized difference = 0.20, and (3) $p = .05$. *n* would be 800 (400 subjects to each group).
- By applying a sample size formula for *n* estimate of an individual group

$$n_i = \frac{[p_A(1 - p_A) + p_B(1 - p_B)]}{(p_A - p_B)^2} \cdot C_{p,power} \tag{10.6}$$

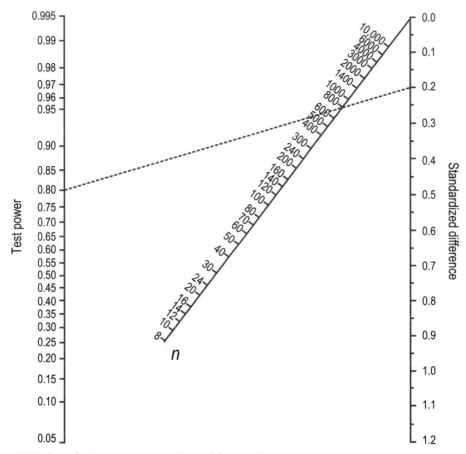

Figure 10.3 Sample size nomogram adjusted for $p < .05$.

where n_i is the n of an individual group, p_A the group A proportion, p_B the group B proportion, $C_{p,\text{power}}$ the constant defined by p and test power. $C_{p,\text{power}}$ can be determined according to Table 10.1:

$$n_i = \frac{[0.42(1 - 0.42) + 0.32(1 - 0.32)]}{(0.42 - 0.32)^2} .7.9 = 356$$

where n_i would be 356 (therefore n would be 712).

Note: sample size nomogram and sample size formula do not necessarily yield the exact same values, assuming their objective is to give n *estimates*.

The above-exemplified values do not represent absolutely "correct" n for reaching reliable results in clinical trials, but only an approximation. For example, if n of 90 is estimated, it is permissible to rule out the need for n of 600, but not for n of 100. Roundups are also possible, for instance, from an estimated n of 178 to n of 180.

10.3 Assessing *n* of a published trial

It is possible to indirectly estimate whether *n* of a published trial was adequate to attain the objective of a study by estimating the respective test power, based on the following rationale: if to determine *n* we first had to choose the test power and then finding the standardized difference now, for determining test power, we must first find the standardized difference—according to published study results—then apply it, the chosen statistical significance level and *n* in the published trial, to the sample size nomogram.

Obs.: the following methods are applicable for equal sized groups.

10.3.1 Studies that analyzed differences between means

For example, in a study with *n* of 200 subjects with COPD (chronic obstructive pulmonary disease), 95 patients were randomized between group A (bronchodilators + nasal O_2) and 105 to group B (only bronchodilators), to compare arterial pO_2 with each modality. Chosen statistical significance level was 0.05. Mean arterial pO_2 in group A was 95 mmHg and mean arterial pO_2 in group B was 93 mmHg, that is, a 2 mmHg difference. According to literature data, the standard deviation for arterial pO_2 in COPD patients treated only with bronchodilators is ± 5 mmHg. Standardized difference is calculated according to formula (10.3).

$$\frac{2}{5} = 0.4$$

Applying data to the sample size nomogram (Fig. 10.4), we can verify that the test power applied in this published trial was 0.80 and that to find such a 2 mmHg difference under a test power or 0.80, *n* of approximately 200 would have been necessary.

10.3.2 Studies that analyzed differences between proportions

For example, in a study with *n* of 390 subjects with moderate bilateral knee osteoarthritis, 189 patients were randomized to group A (physical therapy) and 201 to group B (an OA modifying drug), to compare pain improvement with each modality. Chosen statistical significance level was 0.05. Thirty-eight percent of subjects presented pain improvement in group A, against 29% in group B, that is, a 9% difference. Standardized difference is calculated according to formula (10.5).

$$\frac{0.38 - 0.29}{\sqrt{0.33\,(1 - 0.33)}} \sim 0.2$$

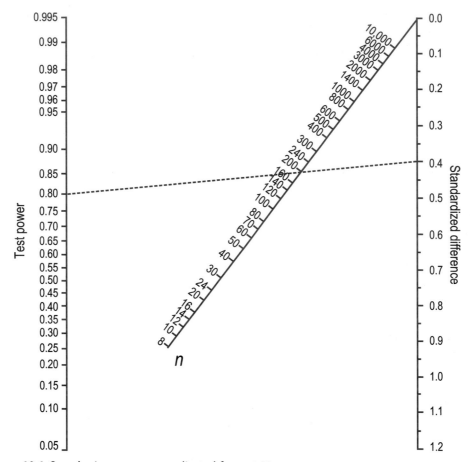

Figure 10.4 Sample size nomogram adjusted for $p < .05$.

Applying data to the sample size nomogram (Fig. 10.5), we can verify that test power applied in this published trial would have been approximately 0.55 (*continuous line*) and that for finding such a 9% difference under a test power or 0.80, *n* of approximately 800 would have been necessary (*dotted line*).

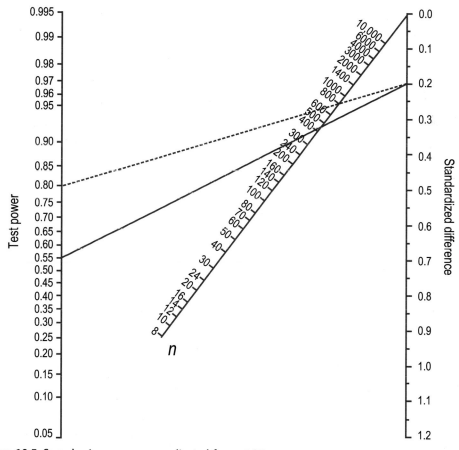

Figure 10.5 Sample size nomogram adjusted for *p* < .05.

CHAPTER 11

Organization of variables and endpoints

The term "variable" refers to any parameter that varies and can be measured (e.g., potassium serum levels, body height, and QRS complex amplitude on EKG). Endpoints are the predictive variables (or efficacy variables) chosen as comparison parameters meant to determine the outcome between the groups of a clinical trial. Variables and endpoints are some of the elements which determine the biostatistical model to be adopted. Variables can be classified as in the following sections.

11.1 Qualitative variables

Qualitative variables do not allow direct assignment of absolute numerical values. An assigned value cannot, in principle, be greater or smaller than the other.

11.1.1 Categorical

Categorical (nominal) variables express characteristics rather than numerical values and do not allow for ordering (e.g., eyes color or type of pain sensation). Nevertheless, to make statistical testing possible, the biostatistician needs to attribute numerical values to the different categories, through appropriate tools:
- Dichotomous variables

    Dichotomous variables admit two mutually exclusive categories (e.g., yes or no, male or female, and life or death).
- Nondichotomous variables

    Nondichotomous variables admit two or more nonmutually exclusive categories (e.g., simple, comminutive, or open fracture; short, moderate, or prolonged sun exposure).

    A numerical value is expected to be attributed to each category of categorical values.

11.1.2 Ordinal

In ordinal variables setting, ordering is admissible, although interval sizes between orders are not quantifiable [e.g., edema intensity (mild, moderate, and severe), social level (high, average, and low)]. Alternatively, categories can be replaced by ranks to facilitate hypothesis testing (*Chapter 14: Hypothesis Testing*). One should be aware that

Practical Biostatistics
DOI: https://doi.org/10.1016/B978-0-323-90102-4.00009-6

biostatistical test results and interpretation will inevitably lead to some imprecision, assuming the initial categorization was elaborated on a subjective basis.

11.2 Quantitative variables

In quantitative variables setting, values are numerically expressed and intervals between these values are equal (e.g., intracranial pressure, body temperature, and number of cells/mm^3). We detail here two types of quantitative variables:

- Discrete

    In discrete (i.e., individual, single) variables setting, only whole values are admissible (e.g., number of pregnancies or convulsion episodes). Discrete variables are generally associated to events counting.

- Continuous

    Continuous (interval) variables refer to whole values as well as their fractions [e.g., age (1 year and 3 months), body weight (48.2 kg)]. Continuous variables are generally associated to some measurement procedure.

    After data collection, variable values will be tabulated according to study design. Clinical research tables are structured as lines, columns, blocks, and repetitions (observations) which correspond to individual subjects, see, for example, in Table 11.1.

Table 11.1 Placebo-controlled efficacy study on an oral hypoglycemic drug.

Treated groups	Tested medications					
	Oral hypoglycemic drug			*Placebo*		
Therapeutic group	Fasting serum glucose (mg/dL)					
Subject 1	102	101	98	134	130	221
Subject 2	95	122	130	131	128	150
Subject 3	109	115	191	409	199	199
Subject 4	99	100	110	97	333	161
Subject 5	77	101	100	102	155	320
Placebo group	Fasting serum glucose (mg/dL)					
Subject 1	99	100	110	131	128	150
Subject 2	102	101	98	102	155	320
Subject 3	77	101	100	97	333	161
Subject 4	95	122	130	409	199	199
Subject 5	109	115	191	134	130	221

Tabulation will be the first step to define the *distribution pattern* of collected variables: *normal* or *nonnormal* (*Chapter 13: Determination of Normality or Nonnormality of Data Distribution*). This will be a defining step regarding the selection of one of the two main biostatistical test groups to be applied: *parametric* (involves normal distribution) or *nonparametric* (involves nonnormal distribution). The best statistical test to determine if $p < \alpha$ will be selected based on this choice.

CHAPTER 12

Measures for results expression of a clinical trial

As previously seen, data frequency distribution from a study can be tabulated. Notwithstanding, assuming it is not practical to express study results in this manner, specific summarizing measures are usually adopted. In normal data distribution studies (Chapter 13: Determination of Normality or Nonnormality of Data Distribution), those data tend to concentrate around a mean an then to disperse bidirectionally. Hence, in this setting, two main types of summarizing parameters are proposed: *central tendency* and *dispersal* measures. As well as summarizing study data, both are useful in estimating the biological qualities of the studied sample.

By convention, the symbols used to express some of central tendency and dispersal measures, are expressed in Greek and English characters for populations and samples, respectively: (1) μ (pronounced as "miu"), mean for population; (2) σ (sigma), standard deviation for population; (3) $\bar{x}$ (pronounced as "x bar"), for samples; and (4) S, standard deviation for samples.

Note: Determination of the parameters presented in this chapter can be performed through widely available online resources. We will detail corresponding calculations in a conventional manner, for didactical purposes.

12.1 Central tendency measures

12.1.1 Mean

Mean (arithmetic mean) ($\bar{x}$ or μ) is the measure that best represents the centrality of a population or sample. This proximity is directly proportional to the following parameters: (1) N or n, respectively; (2) homogeneity of data distribution; and (3) number of observations. It is a useful measure for continuous variables. For example, the results of serum total cholesterol measurements from a 10-patient sample are detailed in Table 12.1 for mean determination.

The mean is determined by the formula:

$$\bar{x} = \Sigma x_i / n \tag{12.1}$$

where $\bar{x}$ is the mean, $\sum$ the summation, x_i the individual variable result, n the number of individuals.

$$\bar{x} = \frac{241 + 190 + 202 + \cdots + 184 + 213 + 236}{10} = 228 \text{ mg/dL}$$

Practical Biostatistics
DOI: https://doi.org/10.1016/B978-0-323-90102-4.00007-2

Table 12.1 Results of serum total cholesterol levels in a sample of 10 patients.

Patients	Serum total cholesterol (mg/dL)
Patient 1	241
Patient 2	190
Patient 3	202
Patient 4	210
Patient 5	299
Patient 6	256
Patient 7	249
Patient 8	184
Patient 9	213
Patient 10	236

Table 12.2 Results of serum total cholesterol levels with two outliers (in bold) ($\bar{x} = 238$).

Patients	Total serum cholesterol (mg/dL)
Patient 1	**110**
Patient 2	190
Patient 3	202
Patient 4	210
Patient 5	299
Patient 6	**492**
Patient 7	249
Patient 8	184
Patient 9	213
Patient 10	236

A limitation of mean is that outliers can significantly deviate it from the value that summarizes the sample or population. Observe a second example based on the same sample, where two outliers replace the values of patients 1 and 6 (Table 12.2).

12.1.2 Median

If the values of a variable from a population or sample are disposed in crescent order, then median will correspond to the value from which half the values are above it and the other half below. For example, a sample of 11 infants with a diagnosis of *Haemophilus influenzae* meningitis was submitted to spinal tap for white cell count in cerebrospinal fluid (Table 12.3).

White cell count results are displayed in a crescent order (Fig. 12.1):

Median is 4500, since there are 5 values above and 5 below it. Its advantages are as follows: (1) relatively protected against outliers (e.g., if the value from patient 9 was 100,000 white cells/mm^3 instead of 5500, median would still remain 4500); (2) useful for observations with a nonnormal distribution (Chapter 13: Determination of Normality or Nonnormality of Data Distribution), for similar reasons; and (3) applicable for discrete as well as ordinal variables. Small possibility for mathematical manipulation is its limitation. For samples with an even number of individuals, median will correspond to the mean of the two central values.

Table 12.3 Results of white cell count in cerebrospinal fluid, in an 11 infants sample.

Patients	White cell count (mm^3)
Patient 1	4600
Patient 2	4000
Patient 3	4600
Patient 4	3500
Patient 5	5000
Patient 6	4250
Patient 7	4550
Patient 8	4500
Patient 9	5500
Patient 10	3600
Patient 11	4200

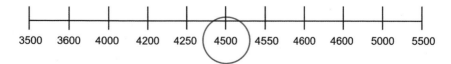

Figure 12.1 Schematic representation of the ordering of Table 12.3 white cell counts.

12.1.3 Mode

Mode corresponds to the value with the largest number of observations in a population or sample. In the example of Section 12.1.2, mode would be 4600. Assuming there may be more than one most frequently observed value, there may be more than one mode. Correspondingly, if there are no values with two or more repetitions, there will be no mode. It is a useful measure for qualitative variables. Mode is a position measure rather than a central tendency measure, given the fact it indicates the point where the largest number of observations is concentrated, which is not necessarily centrally located.

12.2 Dispersal measures

12.2.1 Amplitude

Amplitude corresponds to the difference between the largest and smallest value in a population or sample. In the example depicted in Table 12.1 the amplitude would be 114 mg/dL. Amplitude neither provides information regarding data dispersal nor robustness against outliers.

12.2.2 Variance and standard deviation

As observed in Section 12.1.1, the more disperse the data from a population or sample, the smaller the homogeneity expected from both, from a biological standpoint. One way of measuring this dispersibility is to calculate its mean and then to determine the difference (*deviation*) between each one of its data points and the mean. Let us check the example of Section 12.1.1 again (Table 12.1). Its mean is 228 and deviations are detailed in Table 12.4.

Table 12.4 Results of deviations from the mean of Table 12.1.

Patients	Deviations
Patient 1	$241 - 228 = 13$
Patient 2	$190 - 228 = -38$
Patient 3	$202 - 228 = -26$
Patient 4	$210 - 228 = -18$
Patient 5	$299 - 228 = 71$
Patient 6	$256 - 228 = 28$
Patient 7	$249 - 228 = 21$
Patient 8	$184 - 228 = -44$
Patient 9	$213 - 228 = -15$
Patient 10	$236 - 228 = 8$

Notwithstanding, simply obtaining the deviations of a sample would not suffice to measure its degree of dispersal. To infer it, one must determine the *variance* (σ^2 for populations, S^2 for samples), a measure that condenses all individual deviations, in such a way that either adding or subtracting it from the mean, would provide the range within which most of the data would tend to concentrate and outside of which the remaining data would tend to disperse. Variance is determined by the following formulas for populations and samples, respectively:

$$\sigma^2 = \Sigma(x_i - \bar{x})^2 / N \tag{12.2}$$

$$S^2 = \Sigma(x_i - \bar{x})^2 / n - 1 \tag{12.3}$$

where x_i is the individual variable result, $\bar{x}$ the mean, and N and n are the number of individuals.

The reason why the deviations are squared is that, if otherwise, their sum would equal zero due to the mutual compensation between positive and negative deviations [Table 12.4: $13 + (-38) + (-26) + (-18) + 71 + 28 + 21 + (-44) + (-15) + 8 = 0$]. A numerator equal to 0 would obviously turn the above equations unviable. By squaring the deviations, they are all turned into positive values, allowing calculation. However, squaring the deviations creates a new unit (mg/dL^2), different from the original one (mg/dL), generating interpretation incongruences. Notwithstanding, this can be corrected by calculating *standard deviation* ("standard" meaning representative), through respective formulas:

$$\sigma = \sqrt{\sigma^2} \tag{12.4}$$

$$S = \sqrt{S^2} \tag{12.5}$$

Hence, standard deviation is nothing but the square root of the variance. Let us use the former example for calculating both:

$$S^2 = \frac{(13)^2 + (-38)^2 + (-26)^2 + \cdots + (-44)^2 + (-15)^2 + (8)^2}{10 - 1} = 1233 \ mg/dL^2$$

$$S = \sqrt{1233} = 35 \ mg/dL$$

If one subtracts 35 from 228 ($228 - 35 = 193$) and sum 35 to 228 ($228 + 35 = 263$), an interval representative of the dispersibility of this sample—193 to 263—can be identified, and most of its data is expected to concentrate within it. An alternative way to express this interval is 228 ± 35.[1]

---

[1] From a strict algebraic perspective, variance and standard deviation can only be expressed as positive values; nevertheless, we will express both as negative values also, for didactic purposes.

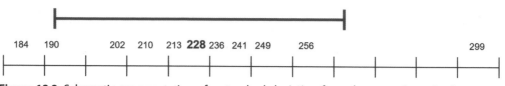

Figure 12.2 Schematic representation of a standard deviation from the mean (superior line represents standard deviation range).

The concept of standard deviation may also be schematically represented. Observe the scale depicted in Fig. 12.2 and notice the tendency of individual values toward the mean.

Classifying the dispersivity expressed by a standard deviation as narrow or large might depend either upon the investigator's interpretation or on the coefficient of variation (CV) (Section 12.2.3). As a general rule, the narrower the standard deviation is, the more homogeneous its corresponding population or sample is expected to be.

12.2.3 Coefficient of variation

Standard deviation by itself does not inform on the relative magnitude of the data of a population or sample. For example, a standard deviation of ± 2 has different meanings in a sample of 10 and in a sample of 100 individuals. Therefore, to make inferences regarding its relative significance, it is necessary to quantify it for population and sample as a percentual representation of dispersal—the CV, through respective formulas:

$$CV = \left(\sigma/\overline{x}\right) \cdot 100\% \tag{12.6}$$

$$CV = \left(S/\overline{x}\right) \cdot 100\% \tag{12.7}$$

where σ is the standard deviation (population), S the standard deviation (sample), and $\overline{x}$ the mean.

By convention, the following CV values are adopted: (1) $<15\%$, low dispersal; (2) $15 < CV < 30\%$, average dispersal; and (3) $>30\%$, high dispersal. CV affords three types of analysis:
- Dispersal estimate of a population or sample

    For example, in a sample of patients with chronic persistent hepatitis, we learn that mean total serum bilirubin levels is 16 mg/dL and standard deviation ± 3 mg/dL. Three is 19% of 16, therefore CV is 19%.
- Comparative dispersal between different samples

    For example, in a sample of patients with chronic persistent hepatitis, we learn that mean total serum bilirubin levels is 16 mg/dL and standard deviation ± 3 mg/dL, therefore CV is 19%. In another sample with the same condition, mean total serum bilirubin levels is also 16 mg/dL and standard deviation ± 5 mg/dL, therefore CV is 31%. So, even though both samples present the same mean, their CVs suggest that they do not evolve similarly.

- Comparative dispersal of variables of different natures in the same sample

  For example, in the very same sample, we have overweight individuals with peripheral insulin resistance, mean body weight of 90 ± 15 kg and mean fasting serum insulin levels of 82 ± 19 mg/dL. Body weight and fasting serum insulin levels CVs are 16% and 21%, respectively. We notice that CVs of both variables are different, which suggests that this sample is more uniform for body weight than for fasting serum insulin levels.

12.2.4 Standard error of the mean

In previous items, we studied mean and standard deviation of a sample and their significance. Nevertheless, we must also take into account that the studied sample has been extracted from a general population and that these statistics may also present some variability in relation to the mean and standard deviation of the general population and remainder samples (Fig. 12.3).

In fact, these samples share similar characteristics since they originate from the same source. On the other hand, they may present different central tendency and dispersal measures among themselves, due to *n* variability and randomness. Therefore it is necessary to know their *error* (error meaning deviation), that is, how much their standard deviations are "bended" by their *n*, before making comparative inferences between central tendency and dispersal measures of two samples and those of the general population and other samples. This is attainable through a dispersal measure named *standard error of the mean*, determined by the following formulas for populations

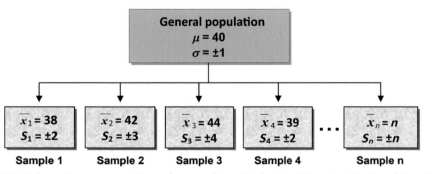

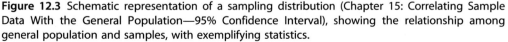

Figure 12.3 Schematic representation of a sampling distribution (Chapter 15: Correlating Sample Data With the General Population—95% Confidence Interval), showing the relationship among general population and samples, with exemplifying statistics.

Table 12.5 *n* and standard deviations from a hypothetical general population and respective samples.

	N/n	σ/S
General population	150	3
Sample 1	50	5
Sample 2	20	2

and samples, respectively:

$$\sigma_\mu = \sigma/\sqrt{N} \tag{12.8}$$

where σ is the population standard deviation, and N the number of individuals in the population.

$$S_{\bar{x}} = S/\sqrt{n} \tag{12.9}$$

where S is the sample standard deviation, and n the number of individuals per sample.

Observe the example (Table 12.5):

$$S_{\bar{x}} = \frac{5}{7} = 0.71 \ (\text{sample 1})$$

$$S_{\bar{x}} = \frac{2}{4.4} = 0.71 \ (\text{sample 2})$$

$$\sigma_\mu = \frac{3}{12.2} = 0.24$$

These figures suggest that, whenever we make comparative inferences, we must take into account that general population *standard error of the mean* tends to be narrower than that of the respective samples.

12.3 Position measures: quantiles

Quantiles can be considered as an extension of the concept of median: while the latter divides the sample or population into two equal halves of observations based on a central value, in quantiles more divisions based on equally distanced values are added, forming new ranges containing each a certain number of observations (not necessarily the same number of observations). Therefore quantiles correspond to a form of sample partitioning into equal frequency ranges, generated by measured variables. Quantiles aim to (1) afford a standardized means to position an individual in a sample, consequently to the corresponding value range; (2) determine an interval established by extreme positions, which encompasses the majority of these individuals, along with related variable ranges; and (3) establish an interval of individuals, along with variable ranges determined by two different positions.

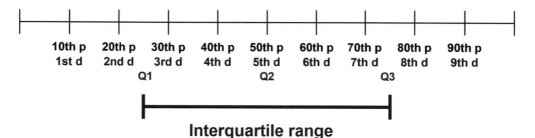

Interquartile range

Figure 12.4 Schematic representation of an interquartile range. *d*, Decile; *p*, percentile; *Q*, quartile.

Table 12.6 Frequency distribution of BMI in a sample of 104 patients.

BMI (kg/m²)	r_j	R_j
>16.5	3	3
16.5-18.5	9	12
18.5-25	13	25
25-30	17	42
30-35	25	67
35-40	28	95
>40	9	104

r_j, Patients per range; R_j, accumulated patients

By convention, there are 99 partitions generated by 100 centiles. Alternatively, centile 10 (or 10th percentile) can be named 1st decile, for example, or centile 25 the 1st quartile (because it corresponds to 1/4 of 100 centiles). The interval between 25th and 75th percentiles corresponds to the so-called interquartile range, which tends to concentrate most part of the individuals of a sample [assuming there is normal distribution (Chapter 13: Determination of Normality or Nonnormality of Data Distribution)] (Fig. 12.4).

Observe this example: BMI was determined in a sample of 104 patients. Results are displayed in a frequency distribution table (Table 12.6).

We wish to know:

- *What BMI range contains centile 90 patient (or the 90th percentile)?*

    First, we must determine who the patient (element) that corresponds to the 90th percentile is. This can be inferred through a simple rule of three formulas:

$$E_c = i(n/100) \tag{12.10}$$

where E_c is the centile element (patient ordinal number), i the proposed centile, n the number of individuals.

$$E_c = 90\left(\frac{104}{100}\right) = 94\text{th element}$$

R_j column shows that 94th patient belongs to the 35–40 kg/m² BMI range. Therefore 35–40 kg/m² BMI is the range that contains 90th percentile.

- At what BMI range will a significant proportion of patients (e.g., 90th percentile) be included?

    According to the former inference, this range corresponds to 35–40 kg/m². This could be interpreted in two ways:

    - 90% of studied patients are included within >16.5–40 kg/m² range (Fig. 12.5).
    - An individual with a BMI above the 35–40 kg/m² range (95th patient on) has a BMI greater than 90% of the remaining individuals in the studied sample.

- *Within which BMI range are most of the patients concentrated?*

    Assuming interquartile range generally comprehends most part of the individuals of a sample, it is expected to correspond to BMI range within which most of the patients are concentrated. Therefore it is necessary to know the patients who correspond to 25th and 75th percentiles:

$$E_c = i_{25}\left(\frac{n}{100}\right) = 26\text{th element}$$

$$E_c = i_{75}\left(\frac{n}{100}\right) = 78\text{th element}$$

The BMI table shows that the interquartile range corresponds to 25–40 kg/m² range, to which most part of patients belong (Fig. 12.6).

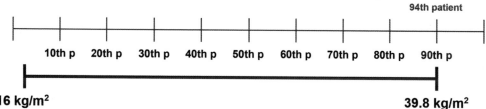

Figure 12.5 >16.5–40 kg/m² range.

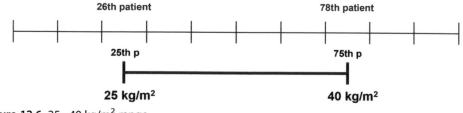

Figure 12.6 25–40 kg/m² range.

An alternative way of expressing these results is a box-and-whiskers plot diagram ("box" corresponding to the interquartile range and "whiskers" to whole observations range) (Fig. 12.7).

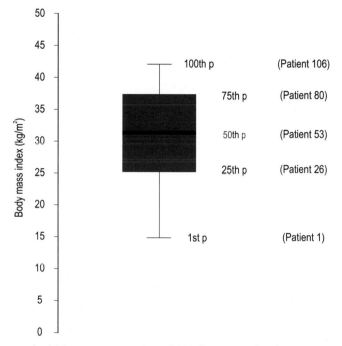

Figure 12.7 A box-and-whiskers representation of BMI frequency distribution.

CHAPTER 13

Determination of normality or nonnormality of data distribution

Note: Determination of the parameters presented in this chapter can be performed through widely available online resources. We will detail corresponding calculations in a conventional manner, for didactical purposes.

Due to their biological nature, it is expected that values collected from a human population follow a typical distribution pattern, characterized by: (1) a central mean which represents most part of these values and (2) the remaining values, which become less frequent as they move away from this mean. For example, a sample of 100 individuals is studied for mean white blood cell count determination (Table 13.1).

Notice that the median—7000 cells/mm^3—concentrates most of the individuals and that the number of individuals gradually decreases as one moves away from it. This type of data distribution is named normal, assuming data distribution in a "normal" population is considered as such. A graphic representation of the distribution of white blood cell counts from this sample is represented in Fig. 13.1.

The type of symmetrical curve generated by a normal distribution, as shown above, is named normal curve. Most of clinical trials data follow a normal distribution pattern, assuming this is the "normal" tendency of biological phenomena.

As shown above, normal curve is prone to present a typical bell-shaped form, determined by a mathematical formula (not shown) which includes mean and standard deviation (SD) parameters. Therefore in order to find out whether a given distribution pattern is normal or not, one necessary step is to plot study data and to verify the curve shape they produce. For this purpose, tabulated data are transposed to a conventional graphic distribution pattern known as Gauss curve. This mathematical resource has the following characteristics:

- their values—named Z-scores—correspond to study data expressed as SD and are determined by a mathematical formula (further in the text)
- Z-scores range from -3 SD to $+3$ SD and are placed in the the horizontal axis
- mean corresponds to a Z-score of 0, with positive SD on the right side and negative SD on the left side
- research subjects are placed in the vertical axis

Practical Biostatistics
DOI: https://doi.org/10.1016/B978-0-323-90102-4.00005-9

Table 13.1 Results of white blood cell counts in a sample of 100 individuals.

	White blood cell counts (cells/mm³)								
	2000	3000	4000	5000	7000	9000	10,000	11,000	12,000
Number of individuals	2	3	5	15	50	15	5	3	2

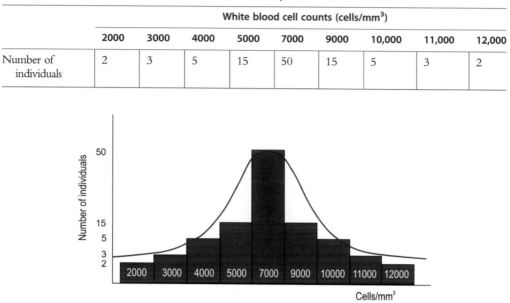

Figure 13.1 A graphic distribution of white blood cell counts in a sample of 100 individuals.

Critical ratio formula is used to convert study data into Z-scores:

$$Z = (x - \mu)/\sigma \tag{13.1}$$

$$Z = (x - \bar{x})/S \tag{13.2}$$

x is a variable value, μ is a mean (populations), $\bar{x}$ is a mean (samples), σ is a standard deviation (populations), and S is a standard deviation (samples).

Gauss curve (Fig. 13.2) presents the following characteristics which make it useful in Biostatistics:

- assuming it would not be practical to perform plottings for every possible data and unit types used in clinical research, Gauss curve could provide an useful standardization
- Gauss curve facilitates comparisons, as well as statistical analyses
- Gauss curve efficiently reproduces the biological characteristics of a population, which tend to present a spiked mean sided by progressively dispersed values (i.e., concentrated within SD range and scarce outside of it)
- Gauss curve graphically expresses the probability of finding a given value in the studied population or sample (the closest to 0, the greater this probability)

In the following example we illustrate how to turn the results of a trial into a Gauss curve: arterial blood oxygen saturation (SO_2) values from a sample of 16 COPD (Chronic Pulmonary Obstructive Disease) patients, admitted in the ICU of a Veterans' hospital, are detailed in Table 13.2.

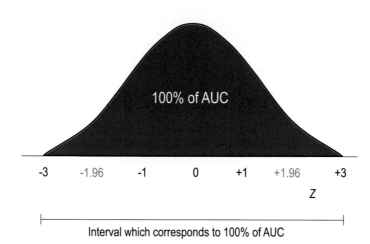

Figure 13.2 The Gauss curve and its main features (AUC—area under the curve).

Table 13.2 Results of arterial blood SO_2 of a sample of 16 patients admitted in the ICU.

Subjects	Arterial blood SO_2 (%)
Patient 1	90
Patient 2	91
Patient 3	89
Patient 4	92
Patient 5	88
Patient 6	93
Patient 7	90
Patient 8	89
Patient 9	91
Patient 10	88
Patient 11	92
Patient 12	90
Patient 13	89
Patient 14	91
Patient 15	90
Patient 16	87

The mean of arterial blood SO_2 in this sample is 90 and SD ± 1.6. Critical ratio formula yields Z-scores, as detailed in Table 13.3.

By tabulating Z-scores according to the number of subjects presenting a specific arterial blood SO_2, we have Table 13.4.

By plotting Z-scores as a Gauss curve, we have Fig. 13.3.

This distribution pattern can be considered as normal.

Table 13.3 Z-scores of the sample of COPD patients from Table 13.2.

Subjects	Arterial blood SO_2 (%)	Z-scores
Patient 1	90	0
Patient 2	91	0.62
Patient 3	89	−0.62
Patient 4	92	1.25
Patient 5	88	−1.25
Patient 6	93	1.87
Patient 7	90	0
Patient 8	89	−0.62
Patient 9	91	0.62
Patient 10	88	−1.25
Patient 11	92	1.25
Patient 12	90	0
Patient 13	89	−0.62
Patient 14	91	0.62
Patient 15	90	0
Patient 16	87	−1.87

Table 13.4 Z-scores correlated to the number of subjects with a specific arterial blood SO_2.

Number of subjects with a specific arterial blood SO_2	Arterial blood SO_2 (%)	Z-scores
4	90	0
3	91	0.62
3	89	−0.62
2	92	1.25
2	88	−1.25
1	93	1.87
1	87	−1.87

In "real world" populations, normal distribution curves do not present themselves as evenly drawn as in the above example. In fact, they usually show nonuniformities which may even cast doubt on their actual normality. Therefore one can infer that the limit between normality and nonnormality may not be well defined, which may render the corresponding decision rather subjective. Some parameters which may aid in this definition are:

- morphology of the curve (asymmetries, skewness, and kurtosis)
- proximity in relation to mean, median, and mode (the closer among themselves, the more "normal" the curve is expected to be)
- a 95% confidence interval (Chapter 15: Correlating Sample Data with the General Population—95% Confidence Interval) that encompasses two SD

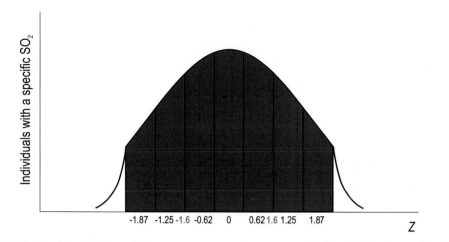

Figure 13.3 Graphic plotting of *Z*-scores corresponding to arterial blood SO_2, from the COPD patients sample. *COPD*, chronic pulmonary obstructive disease.

- a coefficient of variation (Chapter 12: Measures for Results Expression of a Clinical Trial) between 25% and 50%

Consistency or nonconsistency of study data with a normal distribution pattern will determine which might be the most suitable statistical test in order to determine p, or if $p < \alpha$. These tests are represented by two main groups: parametric (consistent with normal distribution) and nonparametric (consistent with nonnormal distribution) (Chapter 14: Hypothesis Testing).

CHAPTER 14

Hypothesis testing

Hypothesis testing consists essentially in comparing central tendency and dispersal measures between two samples[1], in order to test the invesigator's hypothesis. This comparison is performed through special tools named statistical tests, whose aim is to reject or not the null hypothesis (H$_0$ meaning the difference between measures does *not* achieve a statistically significant difference, according to a tabulated critical value). In this chapter the type of measure used to demonstrate how some of these tests work is the mean.

Note: The determination of the parameters presented in this chapter can be performed through widely available online resources. We will detail corresponding calculations in a conventional manner, for didactical purposes.

The parameters usually followed in the process of hypothesis testing are:

- Type of distribution of study data
  1. Parametric tests

     Parametric tests allow a safer H$_0$ rejection in comparison to nonparametric tests (see further). They are named "parametric," for they are based on the gaussian parameters of mean and standard deviation on a normal distribution context (*Chapter 13: Determination of Normality or Nonnormality of Data Distribution*). Some criteria for their applicability are: (1) samples must be independent (*Chapter 9: The Intervention Studies*), (2) dispersal measures between compared samples must be homogeneous, and (3) variables must be quantitative (*Chapter 11: Organization of Variables and Endpoints*).

  2. Nonparametric tests

     They are named "nonparametric," as they are not based on the gaussian parameters of mean and standard deviation. Nonparametric tests are useful whenever parametric tests do not apply. They are statistically weaker and applicable in the following situations: (1) nonnormal data distribution, (2) different conditions prevailing among individuals from the same sample, (3) dispersal measures between compared samples are not homogeneous, (4) qualitative variables involved (specially ordinal variables) (*Chapter 11: Organization of Variables and Endpoints*), and (5) small *n*.

---

[1] In fact, hypothesis testing can involve comparisons of one, two, or more than two samples. In this chapter, only two samples testing will be detailed.

Practical Biostatistics
DOI: https://doi.org/10.1016/B978-0-323-90102-4.00015-1

- Number of samples to be compared: (1) one sample comparison (the sample is compared against the population where it was taken from), (2) two samples comparison, and (3) three or more samples comparison
- Degrees of freedom (ν^2)

   For calculating statistical parameters for populations or samples, one must consider the number of individual variables contained in both. These, of course, "vary," that is, they are "free" to assume different values. Therefore there should be as many "degrees of freedom" as individual variables. For example, if there are 10 individual variables or 10 individuals in a population, there are 10 degrees of freedom in it.

   Due to mathematical concepts not to be discussed here, the number of degrees of freedom coincides with N and $(n-1)$ for population and samples, respectively. In the former example, the number of degrees of freedom is 10. If these 10 individual variables belonged to a sample, then the number of degrees of freedom would be $10 - 1 = 9$.
- Statistical significance level (α) (Chapter 1: Investigator's Hypothesis and Expression of its Corresponding Outcome)

   An α of 0.05 is the usual statistical significance level adopted, and it is usually applied whenever one-sided hypotheses (*Chapter 10: n Estimation and n Assessment of a Published Trial*) are proposed. For example, we wish to verify if A intervention (sample A) or B intervention (sample B) are capable of significantly increasing respective means—$\bar{x}_A$ and $\bar{x}_B$—in comparison with the mean of a reference sample—$\bar{x}_R$ (Fig. 14.1). Most of proposed hypotheses in healthcare sciences yield one-sided results.
- type of relation between samples' individuals (*Chapter 9: The Intervention Studies*): (1) dependent samples and (2) independent samples
- type of variables (*Chapter 11: Organization of Variables and Endpoints*): (1) qualitatite (categorical or ordinal) and (2) quantitative (discrete or continuous)

   Some parametric and nonparametric types of test are exemplified in Table 14.1. Some of the most commonly applied tests will be further detailed.

   For each type of test, there is a corresponding distribution table, which generally associates three parameters: (1) the statistical significance cutoff, as chosen by the investigator (α, generally 0.05), (2) degrees of freedom of studied samples, and (3) samples' n. By associating them, we can find the critical value against which the statistic found by the chosen statistical test must be paired. If test statistic is greater than the critical value, we infer that $P < \alpha$ and H_0 is rejected. If the opposite is inferred, than H_0 can not be rejected.

---

[2] Pronounced as "nu."

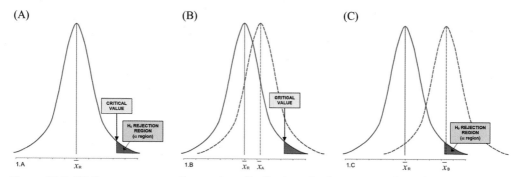

Figure 14.1 (A) Curve corresponding to data distribution of reference sample and $\overline{x}_R$. Critical value (see further) corresponds to the conventional statistical significance cutoff (0.05), beyond which $\overline{x}_A$ and/or $\overline{x}_B$ would enter the H_0 rejection region (α region). (B) Even though $\overline{x}_A$ is greater than $\overline{x}_R$, it does not touch critical value line. Therefore one can *not* reject H_0 (i.e., the hypothesis that there is *no* significant difference between reference sample and sample A means). In other words, there is *no* statistically significant difference between them. (C) $\overline{x}_B$ is greater than $\overline{x}_R$, and it passes the critical value line, entering H_0 rejection region (α region). Therefore one can reject H_0. In other words, there is a statistically significant difference between them.

Table 14.1 Examples of statistical tests classified according to type of distribution of study data and type of comparison.

Parametric		Nonparametric	
— Two samples comparison			
Independent	Dependent	Independent	Dependent
Student's t test	Paired Student's t test	Mann–Whitney test χ^{2a} test	Wilcoxon signed-rank test
Two or more samples comparison			
Independent	Dependent	Independent	Dependent
ANOVA (analysis of variance)	ANOVA	χ^2 test	Cochran test

[a]Pronounced as "chi square."

14.1 Parametric tests for independent and dependent samples

14.1.1 Student's *t* test

Student's *t* test aims to detect differences between means of samples whose data present a normal distribution. A derived statistic named *T* is meant to be compared to a specific value in a t-distribution table—$t_{\nu,\alpha}$—for statistical significance. T is determined by the formula:

$$T = \overline{x}_A - \overline{x}_B / \sqrt{S_c^2(1/n_A + 1/n_B)} \qquad (14.1)$$

Where $\overline{x}_A$ is sample A mean, $\overline{x}_B$ is sample B mean, S_c^2 is combined variance from samples A and B (formula not detailed here), n_A is sample A n, and n_B is sample B n. And degrees of freedom:

$$\nu = (n_A + n_A) - 2 \qquad (14.2)$$

For example, 30 patients with hypertriglyceridemia were randomized between two different therapeutic regimens: (1) sample A—15 patients with lipid lowering diet only and (2) sample B—15 patients with lipid lowering diet plus oral gemfibrozil. Investigator's hypothesis is: could oral gemfibrozil increase triglyceride lowering properties of a lipid lowering diet?

- H_0: diet plus oral gemfibrozil do not lower triglyceride serum levels better than diet only
- H_1: diet plus oral gemfibrozil lower serum triglyceride levels better than diet only
  Mean and standard deviations of both samples are tabulated (Table 14.2).
  T and degrees of freedom are calculated:

$$T = 150 - \frac{35}{\sqrt{33.06(0.13)}} = 1.27$$

$$\nu = (15 + 15) - 2 = 28$$

A t-distribution table is consulted for the critical value corresponding to $t_{28,0.05}$ (part of this table is shown in Table 14.3). We learn that $T = 1.27 < t_{28,0.05} = 1.701$, that is, even though there is a difference between serum triglyceride levels between these samples, it is not significant, from a statistical perspective. In other words, the investigator is not authorized to reject H_0.

Table 14.2 n and statistics of samples A and B.

	Sample A	Sample B
n	15	15
Triglyceride $\overline{x}$	150	135
Standard deviation	33.6	32.4

Table 14.3 Reproduction of part of *t*-distribution table.

DEGREES OF FREEDOM (v)	α		
	0.025	**0.05**	0.10
25	2.060	1.708	1.316
26	2.056	1.706	1.315
27	2.052	1.703	1.314
28	2.048	**1.701**	1.313
29	2.045	1.699	1.311

14.1.2 Paired Student's *t* test

Paired Student's *t* test aims to detect differences between means of the same sample, before and after an intervention (a self-paired sample). A derived statistic T is meant to be compared to a specific value in a *t*-distribution table—$t_{v,\alpha}$—α for statistical significance. T is determined by the formula:

$$T = \frac{\overline{d}}{S_d/\sqrt{n}} \tag{14.3}$$

Where $\overline{d}$ is before and after difference between means, and S_d is standard deviation of the difference (d) between individual values.

And degrees of freedom:

$$v = n - 1 \tag{14.4}$$

For example, the same sample of 10 ER patients with acute respiratory failure under artifical ventilation was tested before and after the use of an aerosolized bronchodilator. Investigator's hypothesis is: could an aerosolized bronchodilator increase tidal volume in a same sample of ER patients with acute respiratory failure under artificial ventilation?

- H_0: artificial ventilation plus aerosolized bronchodilator do not increase tidal volume better than artificial ventilation only, in a same sample of ER patients with acute respiratory failure
- H_1: artificial ventilation plus aerosolized bronchodilator increases tidal volume better than artificial ventilation only, in a sample of ER patients with acute respiratory failure
  Results are detailed in Table 14.4.

$$\overline{x}_{\text{before}} = 697$$

$$\overline{x}_{\text{after}} = 718$$

$$\overline{d} = 718 - 697 = 21$$

Table 14.4 Results of tidal volumes before and after tested treatment (d = difference).

Patient	Before	After	d
1	620	620	0
2	710	720	-10
3	850	870	-20
4	750	765	-15
5	600	625	-25
6	550	590	-40
7	620	620	0
8	690	750	-60
9	790	810	-20
10	79	805	-15

$$S_d = 18.1$$

$$T = \frac{21}{18.1/\sqrt{10}} = 3.75$$

$$v = 10 - 1 = 9$$

A t-distribution table is consulted for the critical value corresponding to $t_{9,0.05}$. We learn that $T = 3.75 > t_{9,0.05} = 2.262$ (the critical value), that is, the difference between tidal volumes before and after aerosolized bronchodilator is statistically significant. In other words, the investigator is authorized to reject H_0 with a 5% probability of being wrong ($P_\alpha < 0.05$).

14.2 Nonparametric tests

Nonparametric tests often involve value ranking, rather than the measured value itself.

14.2.1 For independent samples
14.2.1.1 Mann-Whitney test
Mann-Whitney test aims to detect differences of variable values between two samples through ranking. A derived statistic—U or U′—is meant to be compared to a specific value in a Mann-Whitney distribution table—$U_{\alpha,n1,n2}$ or $U_{\alpha,n2,n1}$—for statistical

significance. If $n_1 > n_2$, then U or U' is compared against $U_{\alpha,n1,n2}$; if $n_2 > n_1$, then against $U_{\alpha,n2,n1}$ (note: comparisons performed in such a way are applicable only to two-sided tests). U and U' are determined by the following formulas, respectively:

$$U = n_1 \times n_2 + \frac{n_1(n_1 + 1)}{2} - R_1 \tag{14.5}$$

$$U' = n_1 \times n_2 - U \tag{14.6}$$

Where n_1 is number of individuals in sample 1, n_2 is number of individuals in sample 2, and R_1 is the sum of sample 1 ranks.

For example, 12 patients with chronic pain were randomized between two different analgesic regimens: (1) sample 1—paracetamol and (2) sample 2—paracetamol plus codein. Investigator's hypothesis is: could codein significantly increase analgesic properties of paracetamol?

- H_0: codein does not increase analgesic properties of paracetamol
- H_1: codein increases analgesic properties of paracetamol

Results are detailed in Tables 14.5 and 14.6.

$$U = 7.5 + \frac{7(7 + 1)}{2} - 61 = 2$$

$$U' = 7.5 - 2 = 33$$

$$U_{\alpha,n1,n2} = U_{0.05,7,5} = 30 \;\; (\text{critical value})$$

Table 14.5 Ranked results for pain levels for sample 1.

Sample 1 patients (n_1)	Pain level (VAS*)	Ranks**
1	43	11th
2	40	12th
3	50	9th
4	45	10th
5	59	7th
6	56	8th
7	79	4th
		$R_1 = 61$

*VAS (visual analogic scale—0−100 mm).
**Ranked from the lowest to the highest variable value.

Table 14.6 Ranked results for pain levels for sample 2.

Sample 2 patients (n_2)	Pain level (VAS)	Ranks**
1	77	5th
2	60	6th
3	90	2nd
4	81	3rd
5	92	1st
		$R_2 = 17$

*VAS (Visual Analogic Scale—0–100 mm).
**Ranked from the lowest to the highest variable value.

A Mann-Whitney distribution table is consulted for the critical value corresponding to $U_{0.05,7,5}$. We learn that $U' = 33 > U_{0.05,7,5} = 30$ (the critical value), that is, the difference between pain levels of these samples is statistically significant. In other words, the investigator is authorized to reject H_0 with a 5% probability of being wrong ($P_\alpha < 0.05$).

14.2.1.2 χ^2 test

The χ^2 test aims to find event rate differences between two or more samples, studied as categorical variables plotted in a two-by-two contigency table (only its applicability between two samples will be depicted). χ^2 is determined by the formula:

$$\chi^2 = \Sigma(\text{observed rate} - \text{expected rate})^2/\text{expected rate} \qquad (14.7)$$

And degrees of freedom:

$$\nu = (\text{lines number} - 1)\ (\text{columns number} - 1) \qquad (14.8)$$

For example, 209 patients in acute lymphoid leukemia first remission were randomized between two different chemotherapy protocols: protocol 1 or protocol 2. After 1 year, remission and relapse rates are determined. Investigator's hypothesis is: would there be significant differences between protocol 1 and protocol 2 remission and relapse rates, after one year of treatment?

- H_0: there are no significant differences between protocol 1 and protocol 2 remission and relapse rates after 1 year of treatment
- H_1: there are significant differences between protocol 1 and protocol 2 remission and relapse rates after 1 year of treatment
  Results are detailed in Table 14.7.

Table 14.7 ALL patients in remission and relapse after one year of treatment, according to protocol type.

	In remission	In relapse	Total
Protocol 1 patients	21 (subgroup A)	51 (subgroup B)	72
Protocol 2 patients	81 (subgroup C)	56 (subgroup D)	137
Total	102	107	209

In order to address this hypothesis, we must first estimate what would be the best possible expected remission and relapse rates, through a simple rule of three:
- Subgroup A
  From a total of 209 patients, 102 remained in remission. From a subtotal of 72 patients treated with protocol 1, 35.1 remissions would therefore be expected.
- Subgroup B
  From a total of 209 patients, 107 relapsed. From a subtotal of 72 patients treated with protocol 1, 36.8 relapses would therefore be expected.
- Subgroup C
  From a total of 209 patients, 102 remained in remission. From a subtotal of 137 patients treated with protocol 2, 66.8 remissions would therefore be expected.
- Subgroup D
  From a total of 209 patients, 107 relapsed. From a subtotal of 137 patients treated with protocol 2, 70.1 relapses would therefore be expected.

$$\chi^2 = \frac{(21-35.1)^2}{35.1} + \cdots + \frac{(56-70.1)^2}{70.1} = 17$$

$$\nu = (2-1)\,(2-1) = 1$$

A χ^2 distribution table is consulted for the critical value (assuming $\alpha = 0.05$ and $\nu = 1$). We learn that $\chi^2 = 17 > 3.841$ (the critical value), that is, the difference between protocol 1 and protocol 2 remission and relapse rates after 1 year is statistically significant. In other words, we are authorized to reject H_0 with a 5% probability of being wrong ($P_\alpha < 0.05$). Since most part of protocol 2 patients remained in remission after 1 year and most part of protocol 1 patients did not, we can infer that protocol 2 yields better therapeutic results than protocol 1 (note: χ^2 test is applicable only if observed rates are greater than 20 and best possible expectable rates are greater than 5).

14.2.2 For dependent samples: Wilcoxon signed-rank test

Wilcoxon signed-rank test aims to detect differences between variables from the same sample before and after an intervention, by calculating the differences between their ranks. A derived statistic named T is meant to be compared to a specific value in T-distribution table—$T_{\alpha,n}$—for statistical significance. Preintervention and postintervention differences are calculated, ranked, and summed up as $T+$ and $T-$.

For example, the same sample of 10 patients with paroxysmal coughing was tested before and after the administration of an experimental cough medicine. Investigator's hypothesis is: could this experimental cough medicine improve paroxysmal coughing according to a cough-related quality-of-life questionnaire (CQLQ), in a same sample of patients?

- H_0: experimental cough medicine does not improve paroxysmal coughing
- H_1: experimental cough medicine improves paroxysmal coughing

    Results are detailed in Table 14.8.

$$T + (\text{all} + \text{signed ranks are summed up}) = 51$$

$$T - (\text{all} - \text{signed ranks are summed up}) = 4$$

Table 14.8 Ranked changes in pretreatment and posttreatment with the experimental cough medicine, with corresponding differences.

Patient	Pretreatment score[a]	Posttreatment score[a]	Difference[b]	Difference rank[c]	Difference signed-rank[d]
1	140	136	4 (4th)	4.5	4.5
2	142	138	4 (5th)	4.5	4.5
3	144	139	5 (8th)	7	7
4	144	147	−3 (3rd)	3	−3
5	146	141	5 (6th)	7	7
6	142	143	−1 (1st)	1	−1
7	150	145	5 (7th)	7	7
8	149	143	6 (9th)	9.5	9.5
9	148	146	2 (2nd)	2	2
10	142	136	6 (10th)	9.5	9.5

[a]Sum of CQLQ parameters.
[b]Ranked from smallest to largest difference.
[c]Calculated according to rank difference, not to the difference itself; whenever ranks coincided with equal differences, corresponding mean was represented by decimal values.
[d]Difference signal is assigned to rank.

A T-distribution table is consulted for the critical value corresponding to $T_{\alpha,n}$. We learn that $T_{0.05,10} = 8$ (the critical value) $> T- = 4$, that is, the difference between pretreatment and posttreatment CQLQ scores is statistically significant (note: $T+$ could also have been used, if it was smaller than $T_{\alpha,n}$ as well). In other words, we are authorized to reject H_0 with a 5% probability of being wrong ($P_\alpha < 0.05$).

CHAPTER 15

Correlating sample data with the general population—95% confidence interval

After determining if there is a statistically significant difference between study group and control group endpoint values, it is necessary to extrapolate corresponding results to the general population. This must be so because it would not be useful for health sciences to collect data that would be applicable only to the studied sample. This correlation is established through the initial determination of the so-called *sample statistics*—mean and standard deviation—, generally extrapolated to the general population as a percentage corresponding to a conventional interval—*confidence interval* (CI)—, found in a normal distribution graph.

One of the principles that allow estimations from a sample to the general population is that *a sample preserves the same random distribution of variable values as the population it was taken from.* For example, suppose we have a 1000 L water tank containing a population of 3000 fish, swimming in random speeds and directions (our variables). Then, we sample 10 L from this tank containing 30 fish in a fishbowl, swimming according to the same random speeds and directions as in the original tank. Even though it is just a sample, it *does* preserve the same randomness of the original population. Assuming it would be cumbersome to determine the speed and direction data of the fish from the whole tank, we need to learn how to "amplify" the randomness of the fishbowl to the tank, through the variable values of the former, to infer the "big picture." This "amplification" is feasible and based on three fundamental concepts—*sampling distribution, central limit theorem,* and *probability theory* (detailed elsewhere).

Sampling distribution consists in distributing a population into samples (each sample containing the same *n*), as in the example depicted in Fig. 15.1.

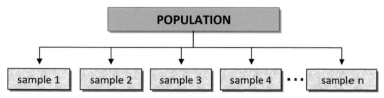

Figure 15.1 *Sampling distribution* of a population.

Practical Biostatistics
DOI: https://doi.org/10.1016/B978-0-323-90102-4.00011-4
125

Sampling distribution is the basis of two population estimation methods: point estimation and the interval estimation.

15.1 Point estimation

Point estimation consists in estimating *population mean* (μ) and *population standard deviation* (σ), based on the so-called *estimators*, that is, *sample mean* ($\mu_{\bar{x}}$) and *standard error of the mean* ($S_{\bar{x}}$), respectively.

15.1.1 Sample mean

Sample mean is determined through the so-called *sampling distribution of the mean*,[1] which consists in distributing the *means of each sample* ($\bar{x}_1$, $\bar{x}_2$, $\bar{x}_3$, etc.) as generated by the *sampling distribution* (Fig. 15.2).

Sample mean corresponds to the mean of the *means of each sample*, as generated by the *sampling distribution of the mean*, and is determined by the formula:

$$\mu_{\bar{x}} = \bar{x}_1 + \bar{x}_2 + \bar{x}_3 + \bar{x}_4 \cdots + \bar{x}_n/\text{number of samples} \tag{15.1}$$

where $\mu_{\bar{x}}$ is the sample mean and $\bar{x}_1, \bar{x}_2, \bar{x}_3\ldots$ the means of each sample.

According to probability theory, one can assume that (1) *sample mean* tends to be equal to the *population mean*, (2) it is possible to estimate the *population mean* if one knows *n* of a single sample and the mean of this sample, and (3) as greater sample *n*, as closer to the *population mean* the *sample mean* is expected to be. According to central limit theorem, one can assume that the *means of each sample* as plotted in a graph tend to appear as a normal distribution curve. The shape of this curve depends on two elements: (1) number of *means of each sample* (as many *means of each sample* we plot, as much "normal" this curve would appear) and (2) sample's *n* (as many individuals per sample, as narrower this curve would appear).

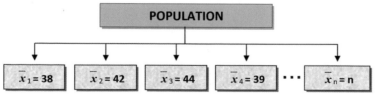

Figure 15.2 *Sampling distribution of the mean* of Fig. 15.1 with exemplifying figures. It is assumed that each sample is composed of a fixed *n*.

---

[1] Actually sampling distribution of the means, for it refers to two or more means.

15.1.2 Standard error of the mean

Standard error ($\sigma_{\bar{x}}$) is an estimation of the standard deviation of the *means of each sample* as generated by the *sampling distribution of the mean*. It is determined by the formula:

$$\sigma_{\bar{x}} = \sigma / \sqrt{n} \qquad (15.2)$$

where $\sigma_{\bar{x}}$ is the standard error, σ the population standard deviation, and n the number of individuals in the sample.

Two points are inferable from this formula:

- The wider the *population standard deviation*, the wider the *standard error*, meaning the *standard error* tends to follow the *population standard deviation*.
- The greater the sample's n, the narrower the *standard error*, meaning that the following differences tend to be smaller: (1) between *standard error* and *population standard deviation* and (2) between *sample mean* and *population mean*.

The limitation of *standard error* is that, in practice, we seldom know how much *population standard deviation* is. Through *standard error of the mean* (Chapter 12: Measures for Results Expression of a Clinical Trial) it is possible to estimate the *standard deviation* using the *sample standard deviation* (*S*), instead of the *population standard deviation*. *Standard error of the mean* is determined by the formula:

$$S_{\bar{x}} = S / \sqrt{n} \qquad (15.3)$$

where $S_{\bar{x}}$ is the standard error of the mean, S the sample standard deviation, and n the number of individuals in the sample.

Two points are inferable from this formula:

- The wider the *sample standard deviation*, the wider the *standard error of the mean*, meaning the *standard error of the mean* tends to follow the *sample standard deviation*.
- The greater sample's n, the narrower the *standard error of the mean*, with an expectation of a smaller difference between the *standard error of the mean* and the *standard deviation*.

In *standard error of the mean* formula the variable *sample standard deviation* has to be estimated by the statistician. However, assuming we are dealing with samples, not with populations (as in *standard error*), this would be less of a problematic task. Based on the above assumptions and on the probability theory, it is therefore possible to propose the *standard error of the mean* as an estimator for *population standard deviation*.

In summary, by determining *sample mean* and *standard error of the mean*, it is possible to estimate the *population mean* and *population standard deviation*, respectively, on a probability basis (complete methodology detailed elsewhere).

15.2 Interval estimation (95% confidence interval)

Even though point estimation provides a solid probability basis for sample to population extrapolations, a precise coincidence between a sample estimator with corresponding population parameter would be an unlikely event, at least in clinical research. The idea of interval estimation is to widen the probability that the statistic found in the studied sample might coincide with the parameter of the general population, by building a conventional interval siding this statistic. Let us take Fig. 13.2 from Chapter 13, Determination of Normality or Nonnormality of Data Distribution, as an example and assume that it derives from a hypothetical original sample (Fig. 15.3).

Based on the above graph, the following points can be inferred:

- The closer to the mean a given sample value is, the greater the probability for its detection in the general population.
- If one assumes an α of 5% (Chapter 1: Investigator's Hypothesis and Expression of Its Corresponding Outcome) regarding the hypothesis that *all values in a sample* can *be found in the general population* (i.e., a 5% probability this hypothesis is *not* true), by inference there would remain a *95% confidence* that such a hypothesis *is* true. In other words, there would be a 95% probability in finding a given sample value in the general population and a 5% probability in *not* finding it. This inference can be expressed by the formula:

$$100\% - \alpha = 95\% \tag{15.4}$$

Hence, 95% would correspond to a conventional *confidence level*. Assuming that the probability of finding a sample value in the general population is inversely proportional to its distance from the *sample mean*, then the values corresponding to α are supposed to be found in the upper and lower extremities of sample's values range. The graphic representation of this inference is depicted in Fig. 15.4.

By convention, we put our "confidence" that the statistic found under the sample's 95% area under the curve (AUC) might coincide with the value of the corresponding parameter, eventually found in the general population. According to the Gauss curve, 95% of AUC correspond to the interval between Z-scores, -1.96 SD (standard deviation) to $+1.96$ SD. Hence, this interval is named *95%* confidence interval (95% CI).

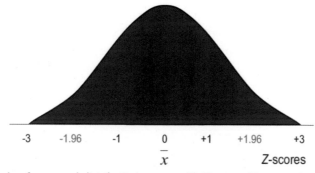

Figure 15.3 Example of a normal distribution curve, with Z-scores ($\bar{x}$ = mean).

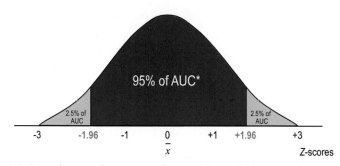

Figure 15.4 The AUC* and interval corresponding to the probability of finding a sample statistic (or endpoint value) in the general population. If we assume that the probability of 95% in finding a sample value in the general population is two-tailed and symmetrical, then we must also distribute α symmetrically, that is, $\alpha/2 = 2.5\%$ per tail. *AUC*, Area under the curve.

$- 1.96$ and $+1.96$ SD correspond to *inferior* and *superior confidence limits* of a sample, respectively. To determine them, we can apply the *confidence limits formula*:

$$x_S = \bar{x} + 1.96 \cdot S_{\bar{x}} \tag{15.5}$$

$$x_i = \bar{x} - 1.96 \cdot S_{\bar{x}} \tag{15.6}$$

where x_S is the *superior confidence limit*, x_i the *inferior confidence limit*, $\bar{x}$ the *mean of the sample*, and $S_{\bar{x}}$ the *standard error of the mean* $= S/\sqrt{n}$ ($S =$ *sample standard deviation*; $n =$ number of individuals in the sample).

Let us check two examples, taken from two different samples of mild renal failure patients, containing 32 individuals each:

- Sample A

    Mean of the sample is equal to 80 mL/min of creatinine clearance and *sample standard deviation* is equal to ± 51. By applying the *confidence limits formula*, we have:

$$x_i = 80 - 1.96 \left(\frac{51}{\sqrt{32}} \right) = 62.0 \text{ mL/min}$$

$$x_S = 80 + 1.96 \left(\frac{51}{\sqrt{32}} \right) = 98.0 \text{ mL/min}$$

In conclusion, sample A has a 95% CI of 62.0—98.0 mL/min. That means that one has a 95% confidence that a randomly selected individual from a population of patients with a clinical picture of mild renal failure might present a creatinine clearance value within this range (as close to *mean of the sample*—80 mL/min—as great the probability).

- Sample B

    Mean of the sample is equal to 80 mL/min of creatinine clearance and *sample standard deviation* is equal to ± 34. By applying the *confidence limits formula*, we have:

$$x_i = 80 - 1.96\left(\frac{34}{\sqrt{32}}\right) = 68.0 \text{ mL/min}$$

$$x_S = 80 + 1.96\left(\frac{34}{\sqrt{32}}\right) = 92.0 \text{ mL/min}$$

In conclusion, sample B has a 95% CI of 68.0–92.0 mL/min. That means that one has a 95% confidence that a randomly selected individual from a population of patients with a clinical picture of mild renal failure, might present a creatinine clearance value within this range (as close to *mean of the sample*—80 mL/min—as great the probability).

By graphically plotting the data from sample A and sample B, we have Fig. 15.5.

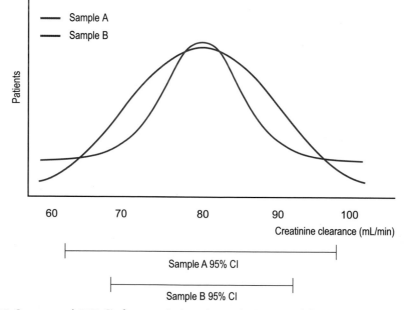

Figure 15.5 Superposed 95% CIs for sample A and sample B. *CI*, Confidence interval.

So, if we have two 95% CIs, on which one can we put our "confidence" the most? The answer depends on the objective of the study. Let us analyze two possible interpretations inferable from each sample:

- Sample A

    Since it has a more heterogenous population from a biological standpoint, it generates a wider standard deviation and consequently a wider 95% CI. This sample represents the biological reality of creatinine clearance in the context of mild renal failure more poorly, since its population is so heterogeneous. On the other hand, corresponding 95% CI can be extrapolated to the general population of patients with mild renal failure more promptly, assuming such a heterogeneity is expected to be found in a "real world" context.

- Sample B

    Since it has a more homogeneous population from a biological standpoint, it generates a narrower standard deviation and consequently a narrower 95% CI. This sample better represents the biological reality of creatinine clearance in the context of mild renal failure, since its population is so homogeneous. On the other hand, corresponding 95% CI cannot be extrapolated to the general population of patients with mild renal failure so promptly, assuming such a homogeneity is not expected to be found in a "real world" context.

Summarizing the steps

All contents described in Parts 1 and 3 can be summarized in the following steps. The investigator:

Step 1: Formulates a hypothesis

Step 2: Selects the most adequate study type for addressing his/her hypothesis

Step 3: Estimates *n*

Step 4: Determines and organizes the variables and endpoints of the study

Step 5: Determines collected data normality or nonnormality

Step 6: Calculates the summarizing mathematical parameters relatively to study data

Step 7: Selects the best statistical test type to either refute or confirm the investigator's hypothesis

Step 8: Correlates the data of the sample with the general population—95% CI is determined

The above sequence is not suitable to address all types of investigator's hypotheses. For different types of investigation lines, we propose other resources, to be detailed in Part 5.

Bibliography

Suggested reading (Part 4)

Altman, D.G., 2010. Statistics and ethics in medical research. III. How large a sample? Br. Med. 281, 1336–1338.

Canadian Medical Association Journal, 2017. Basic statistics for clinicians. www.cmaj.ca. (Accessed 4 May 2017).

Centre for Evidence Based Medicine, 2016. Department of Medicine, Toronto General Hospital. Statistics for the clinic. www.cebm.utoronto.ca (Accessed 3 May 2016).

Drummond, J.P., Silva, E., 2012. Medicina Baseada em Evidências—Novo Paradigma Assistencial e Pedagógico. Atheneu, Rio de Janeiro (Chapter 4).

Estrela, C., 2001. Metodologia Científica—Ensino e Pesquisa em Odontologia, first ed. Editora Artes Médicas, Porto Alegre (Chapter 7).

Everitt, B., 2006. Medical Statistics From A to Z. A Guide for Clinicians and Medical Students, second ed. Cambridge University Press, London.

Everitt, B.S., 2008. The Cambridge Dictionary of Statistics in the Medical Sciences. Cambridge, London.

Everitt, B.S., et al., 2005. Encyclopaedic Companion to Medical Statistics. Hodder Arnold, London.

Hulley, S.B., et al., 2001. Designing Clinical Research: An Epidemiological Approach, second edition Lippincott Williams & Wilkins, Philadelphia, PA.

Medronho, R.A., et al., 2009. Epidemiologia, second ed. Atheneu, Rio de Janeiro.

Merriam-Webster Online Dictionary. www.merriam-webster.com.

Neto PLOC, 1977. Estatística. Editora Edgard Blücher, São Paulo.

Oliveira, G.G., 2006. Ensaios Clínicos. Princípios e Prática. Editora Anvisa, Brasilia.

Sackett, D.L., et al., 2001. Evidence-Based Medicine: How to Practice and Teach EBM, second ed. Elsevier Health Sciences, Amsterdam.

Schmuller, J., 2009. Statistical Analysis With Excel for Dummies, second ed. Wiley Publishing, Hoboken, NJ.

Whitley, E., Ball, J., 2002. Statistics review 4: sample size calculations. Crit. Care 6, 335–341.

PART 5

Additional concepts in biostatistics

The objective of Part V is to expand the reader's proficiency in biostatistics, by detailing other approaches used in clinical research. The chapters were developed independently and can be studied separately.

CHAPTER 16

Individual and collective benefit and risk indexes inferable from intervention studies

It is possible to extend the usefulness of the data from a clinical study beyond statistical significance (*p*) and 95% confidence interval, through benefit and risk indexes. They derive specially useful tools in the clinical setting and will be further detailed based on the results of an actual clinical trial: subcutaneous ondaparin for the prevention of venous thromboembolism in ICU admitted medical patients—a randomized placebo-controlled trial [*n* = 644 (ondaparin group 321 patients, placebo group 323 patients)]. Based on its results, it is possible to tabulate undesired event [deep venous thrombosis (DVT)] and adverse reaction rates, associated to the tested therapy (Tables 16.1 and 16.2, respectively). Some hypothetical rates will also be useful for detailing additional concepts.

The following indexes can be inferred from the above mentioned tables.

16.1 Treatment effect indexes

Treatment effect indexes refer to expected collective occurrence rates and are not applicable in the setting of individual patients.

16.1.1 Risk indexes

16.1.1.1 Basal risk
Basal risk (BR) refers to the degree of risk in the control group, regarding undesired events as well as adverse reactions (in our example 10.5% and 3.2%, respectively).

16.1.1.2 Relative risk
Relative risk (RR) determines the degree of risk for the undesired event in the subgroup which presented it (experimental group), relatively to BR (Fig. 16.1). It is determined by the formula:

$$\frac{\text{UEE} \cdot 100}{\text{UEC}} \tag{16.1}$$

Table 16.1 Undesired event rates (deep venous thrombosis) from ondaparin clinical trial.

	UEC (%)	UEE (%)
Undesired event real rate	10.5	5.6
Undesired event *hypothetical* rate	20	4
Extremely low undesired event *hypothetical* rate	0.1	0.05
Extremely high undesired event *hypothetical* rate	95	90

UEC, undesired event rate in the control group; UEE, undesired event rate in the experimental group.

Table 16.2 Adverse reaction rates from ondaparin clinical trial (some data have been changed for didactic purposes).

	Adverse reaction rates (%)
ARC	1.2
ARE	3.2

ARC, adverse reaction rate in the control group; ARE, adverse reaction rate in the experimental group.

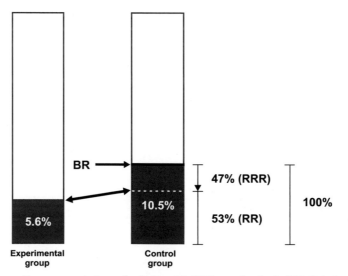

Figure 16.1 Schematic representation of relative risk (RRR: see further). *RRR*, Relative risk reduction; *RR*, Relative risk.

$$\frac{5.6 \cdot 100}{10.5} = 53\%$$

The above result means that experimental group had a 53% degree of risk of presenting the undesired event, relatively to the control group. In other words, if the patients of the control group had received ondaparin, only 53% of them would have presented DVT. This index is limited by its low capacity of expressing valid results, whenever extremely low or high frequencies are reported (Table 16.1):

• RR calculated for extremely low hypothetical rates

$$RR = \frac{0.05 \cdot 100}{0.1} = 50\%$$

Even though the number of patients represented by these hypothetical frequencies is neglectful (0.3 patient on placebo and 0.1 patient on ondaparin), RR reduction (RRR) would be relatively high. Ondaparin association with a decrease in DVT frequency would be, therefore, overestimated. In this setting, odds ratio (OR) (*Section 16.1.1.7*) would be a more suitable index.

• RR calculated for extremely high hypothetical rates (Fig. 16.2)

$$RR = \frac{90 \cdot 100}{95} = 95\%$$

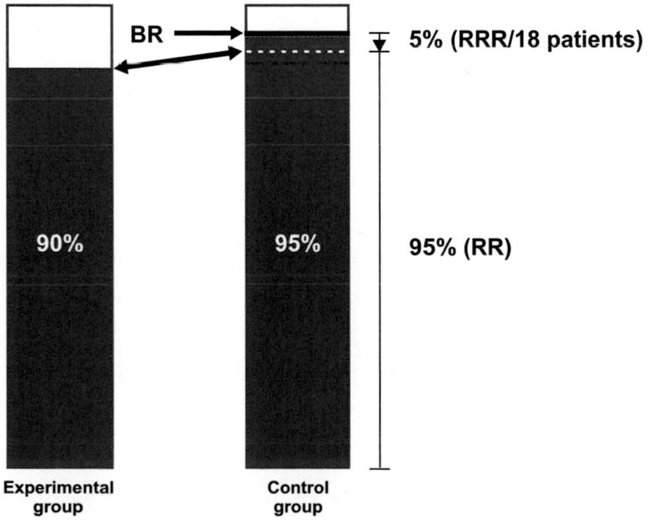

Figure 16.2 Schematic representation of relative risk in extremely high hypothetical rates setting. *RRR*, Relative risk reduction; *RR*, Relative risk.

The important difference between the number of patients of the subgroup that presented DVT (control group) relatively to the opposite subgroup (experimental group) (307 patients − 289 patients = *18 patients*) would point to a considerable ondaparin efficacy. However, a 95% RR suggests that this efficacy would be, nevertheless, exaggerated.

16.1.1.3 Relative risk reduction

RRR determines the degree of risk reduction for the undesired event in the subgroup which presented it (experimental group), relatively to the opposite subgroup (control group). It is determined by the formula:

$$\frac{UEE - UEC}{UEC} \tag{16.2}$$

$$\frac{5.6 - 10.5}{10.5} = 0.46 \text{ (or } 46\%)$$

Observe the first histogram from Section 16.1.1.2.

The above result means that ondaparin was associated to a 46% DVT risk reduction in the experimental group, relatively to the control group. In other words, if the patients of the control group which presented the undesired event had received ondaparin, they would have had a 46% lesser probability of presenting DVT.

Similarly to RR, RRR is limited by the reduction in its capacity of expressing valid results, whenever extremely low or high frequencies are reported (Table 16.1):

• RRR calculated for extremely low hypothetical rates

$$\frac{0.05 - 0.1}{0.1} = 0.5 \text{ (or } 50\%)$$

Even though the number of patients represented by these hypothetical frequencies is neglectful (0.006 patient on placebo and 0.003 patient on fondaparinux), RRR reached 50%. Association of ondaparin with DVT frequency decrease would be, therefore, overestimated. In this setting, OR (*Section 16.1.1.7*) would be a more suitable index.

• RRR calculated for extremely high hypothetical rates (observe the second histogram from *Section 16.1.1.2*)

$$\frac{90 - 95}{95} = -0.05 \text{ (or } 5\%)$$

The significant difference between the number of patients of the subgroup that presented DVT (control group) relatively to the opposite subgroup (experimental group) (307 patients − 289 patients = *18 patients*) would point to a considerable ondaparin efficacy. However, RRR was only 5%. Ondaparin efficacy in decreasing DVT frequency would be, therefore, underestimated.

16.1.1.4 Absolute risk reduction

Absolute risk reduction (ARR) determines the degree of risk reduction for the undesired event between control (BR—*Section 16.1.1.1*) and experimental groups (Fig. 16.3). It is calculated by the simple arithmetic difference between UEC and UEE:

$$10.5 - 5.6 \ = \ \sim 5\%$$

The above result means that if ondaparin had been used in all of the patients of the control group, they would have had a 5% lesser probability of presenting DVT.

Differences between ARR and RRR (*Section 16.1.1.3*) are:

- Greater proximity to "real world" scenarios

  While RRR focuses only on the subgroups which presented the undesired event, ARR encompasses the whole group (the subgroup which presented and the subgroup which did not present the undesired event, from both groups). This assumption incurs in considering not only the risk variability among those who presented the undesired event but BR (*Section 16.1.1.1*) variability as well. This is a situation closer to decision-making processes in the "real world," assuming one cannot know beforehand if a procedure might direct a patient either to the subgroup which will present or to the subgroup which will not present the undesired event.

- Protection against distorted results based on extremely low values

  Observe ARR calculated for extremely low hypothetical rates (Table 16.1).

$$0.1 - 0.05 \ = \ \sim 0.05\%$$

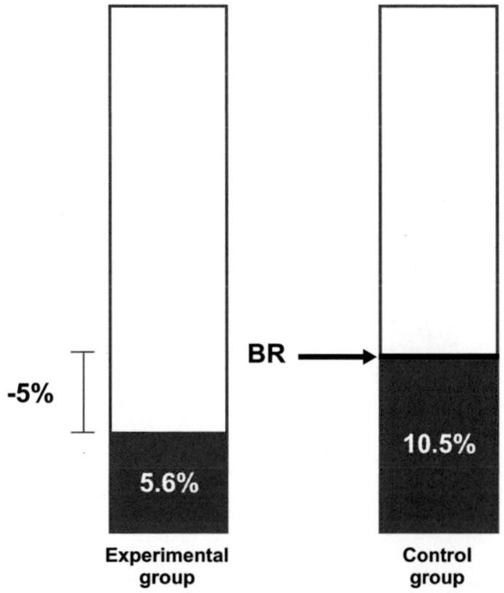

Figure 16.3 Schematic representation of absolute risk reduction. *BR,* Basal risk.

It is obvious that there is no significant ARR in the above example, even though corresponding RRR (*Section 16.1.1.3*—RRR calculated for extremely low hypothetical rates) suggests the opposite.

A 5% RRR index based on real rates is consistent with what is inferable from a 46% RRR, that is, that there *is* a risk reduction in DVT among older medical patients under ondaparin, relatively to placebo. Therefore one can consider the possibility of spurious interpretations due to mathematical nuances as unlikely.

One ARR limitation is its decreased capacity of BR variability expression, in situations where RR (*Section 16.1.1.3*) and RRR variability prevail, along with a constant ARR. Observe the hypothetical evolving example, detailed in Fig. 16.4.

RRR and RR clearly better expressed evolutive BR variability than ARR, which remained equal (10%) from Time 1 to Time 2.

16.1.1.5 Relative risk increase

RR increase (RRI) determines the degree of risk increase for an adverse reaction in the subgroup which presented it (experimental group), relatively to the opposite subgroup (control group). It is determined by the formula:

$$\frac{ARE - ARC}{ARC} \tag{16.3}$$

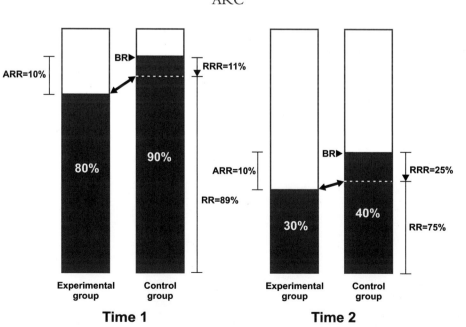

Figure 16.4 Schematic representation of an evolving sample (Time 1 and Time 2), where BR seems to express different meanings. *ARR*, Absolute risk reduction; *BR*, Basal risk; *RR*, Relative risk.

$$\frac{3.2 - 1.2}{1.2} = 1.6 \ (\text{or} + 60\%)$$

The above result means that if ondaparin had been used in all patients from the subgroup that presented hemorrhage (control group), they would have had a 60% greater probability of presenting this adverse reaction.

16.1.1.6 Absolute risk increase

Absolute risk increase (ARI) determines the degree of risk increase for an adverse reaction in an experimental group. Calculated by the simple arithmetic difference between ARE and ARC:

$$3.2 - 1.2 = 2\%$$

The above result means that, if ondaparin had been used in *all patients of the control group*, they would have had a 2% greater probability of presenting hemorrhage episodes.

While RRI (*Section 16.1.1.5*) focuses only on the subgroups which presented the adverse reaction, ARI encompasses the whole group (the subgroup which presented and the subgroup which did not present the adverse reaction, from both groups). This assumption incurs in considering not only the risk variability among those who presented the adverse reaction but BR (*Section 16.1.1.1*) variability as well. This is a situation closer to decision-making processes in the "real world," assuming one cannot know beforehand if a procedure might either direct to the subgroup which will present or to the subgroup which will not present the adverse reaction.

16.1.1.7 Odds ratio

In clinical trial setting, OR expresses the odds for the occurrence of an undesired event in the experimental group, relatively to the same odds in the control group. It is determined by the formula:

$$\frac{\text{UEE}/(100 - \text{UEE})}{\text{UEC}/(100 - \text{UEC})} \tag{16.4}$$

Let us check what would be the ratio between the undesired event odds in the ondaparin group and the same odds in the placebo group:

$$\frac{0.06}{0.12} = 0.5$$

The above result means that the experimental group had a 0.5:1 ratio of presenting an undesired event, relatively to control group. OR express essentially the same as RR (*Section 16.1.1.2*), with the following differences:

- Internists and surgeons expect to obtain immediate applicable information from intervention studies and the information detailed here is often better captured in the practice setting as risk rather than as odds: to inform that ondaparin affords a 53% reduction in DVT risk, is generally better understood than stating that ondaparin presents a 0.5:1 odds for this undesired event.
- RR informs more clearly on no treatment effect than OR: not administering ondaparin would increase DVT risk by 47% (100% − 53%).
- OR informs on the undesired event occurrence odds in the groups as a whole—a more appealing information in the epidemiological setting, while RR focuses on how much BR (*Section 16.1.1.1*) is prone to variation between subgroups—a more appealing information in the practice setting.
- OR is a more suitable resource to estimate the degree of association between an exposure factor and risk of an undesired event, for low rate events.

OR is specially useful in case-control observational studies (*Chapter 6: Basic Concepts in Observational Studies*) and in metaanalyses (*Chapter 18: Systematic Reviews and Metaanalyses*).

16.1.2 Benefit indexes

16.1.2.1 Absolute benefit increase

Absolute benefit increase determines the degree-of-benefit increase in the experimental group (benefit defined as the nonoccurrence of the undesired event), which is calculated by the simple arithmetic difference between the rate of the benefited subgroup of the control group and the rate of the benefited subgroup of the experimental group:

$$94.4 - 89.5\% = \sim 5\%$$

The above result means that ondaparin use by the experimental group was associated to a 5% probability increase in *not* presenting DVT.

16.1.2.2 Relative benefit increase

Relative benefit increase determines the degree-of-benefit increase in the benefited subgroup of the experimental group, relatively to the benefited subgroup of the control group (benefit defined as the nonoccurrence of the undesired event). It is determined by the formula:

$$\frac{A - B}{B} \tag{16.5}$$

where A is the rate of the benefited subgroup of the control group and B the rate of the benefited subgroup of the experimental group:

$$\frac{94.4 - 89.5}{89.5} = 0.05 \text{ (or 5\%)}$$

The above result means that ondaparin use by the benefited subgroup of the experimental group was associated to a 5% probability for this subgroup in *not* presenting DVT, relatively to the benefited subgroup of the control group.

16.1.3 Number needed to treat

Number needed to treat (NNT) corresponds to the number of individuals that must be treated, so that one subject is benefited by the treatment. It is determined by the formula:

$$\frac{1}{\text{ARR}} \qquad\qquad (16.6)$$

$$\frac{1}{5\ *} = 0.20 = 20$$

The above result means that, if ondaparin had been used in the control group, it would have been necessary to treat 20 patients so that one subject would *not* present DVT. NNT corresponds to ARR (*Section 16.1.1.4*), expressed in a different manner.

Obviously, changes in the elements which determine ARR—UEE and UEC—will influence NNT. As a general rule, NNT changes inversely with BR (*Section 16.1.1.1*), that is, the higher BR (therefore UEC), the lower NNT is expected to be. Observe NNT
calculated for undesired event hypothetical rate (Table 16.1):

$$\frac{1}{20 - 4} = 0.20 = 20$$

$$\frac{1}{16} = 0.06 = 6$$

A greater ARR difference (16%—5%*) might represent a greater therapeutic efficacy in terms of undesired event avoidance. As such, a smaller number of patients to treat would be necessary, so that one of them would be benefited. It is inferable that ondaparin administration would be more advantageous in the latter situation than in the former.

NNT's main usefulness is to make ARR data sound more practical to physicians and comprehensible for patients. Its interpretation must be performed based on the physician's own practice experience and on NNTs established for other treatment modalities related to the case. NNT is not as statistically as sound as ARR.

Other NNT examples are:

- intensive insulin regimen over 6.5 years for diabetic neuropathy prevention: 15;
- streptokinase administration, followed by daily aspirin over 5 weeks for acute myocardial infarction death prevention: 29; and
- antihypertensive medication for 5.5 years for acute myocardial infarction, stroke, and death prevention: 128.

16.1.4 Number needed to harm

Number needed to harm (NNH) corresponds to the number of individuals that must be treated, so that one of them presents an adverse reaction accountable to the treatment. It is determined by the formula:

$$\frac{1}{ARI} \tag{16.7}$$

$$\frac{1}{2} = 0.5 = 50$$

The above result means that, if the control group had used ondaparin, it would have been necessary to treat 50 patients, so that one would present an adverse reaction accountable to the treatment. NNH corresponds to ARI (*Section 16.1.1.6*), expressed in a different manner.

NNH's main usefulness is to make ARI data sound more practical to physicians and comprehensible for patients. Its interpretation must be performed based on the physician's own practice experience and on NNHs established for other treatment modalities related to the case.

16.1.5 Likelihood of being helped versus being harmed

Likelihood of being helped versus being harmed (LHH) is an aggregation ratio that takes NNT and NNH into account:

$$\frac{1}{NNT} : \frac{1}{NNH} \tag{16.8}$$

$$\frac{1}{20} : \frac{1}{50} \rightarrow 3:1$$

The above result means that ondaparin treatment has a probability of 3 of benefiting study patients, against a probability of 1 of harming them.

16.2 Clinical decision analysis indexes

Clinical decision analysis indexes refer to expected occurrence rates related to a specific patient in the practice setting, based on clinical studies results or empirical estimates. Assuming that in the practice setting, one must take into consideration the potential benefits and risks of a treatment modality altogether, these indexes shall take both into account. There are three ways of expressing this type of information.

16.2.1 Patient-specific number needed to treat

Suppose it is necessary to know NNT for a specific patient, in such a way that the least possible degree of risk is expected from the considered therapeutic modality. This index can be calculated using two formulas.

16.2.1.1 First formula

$$\frac{1}{PEER \times RRR} \tag{16.9}$$

where PEER is the patient expected event rate and RRR the relative risk reduction.

PEER and RRR (*Section 16.1.1.3*) can be taken from literary sources which describe a control group consistent with the specific patient. By applying ondaparin study as an example [PEER corresponding to UEC (Table 16.1)]:

$$\frac{1}{10.5 \times 46} = 0.002 = 20$$

The above result means that it would be necessary to treat 20 patients in the same way as the specific patient to obtain a positive result, expecting the least possible degree of risk.

16.2.1.2 Second formula

$$NNT/f_t \tag{16.10}$$

where NNT is the number needed to treat and f_t the fraction$_{treatment}$.

NNT can be taken from literary sources [ondaparin study, for instance (*Section 16.1.3*)]. f_t can be taken from two sources:
- Empirical estimates
     One estimates that the specific patient, if left untreated, would have a doubled risk of presenting the undesired event, relatively to a treated patient.
- Literary sources which present control and experimental groups consistent with the case of the specific patient

In the ondaparin study, patients from the subgroup which presented the undesired event (control group) showed a doubled risk for DVT, relatively to patients from the opposite subgroup (experimental group):

$$\frac{20}{2} = 10$$

The above result means that it would be necessary to treat 10 patients in the same way as the specific patient to obtain a positive result, expecting the least possible degree of risk.

Both formulas for patient-specific NNT determination express essentially the same information. The differences are as follows: (1) the second formula is easier to apply and (2) it allows us to use data learned from our own clinical experience.

16.2.2 Patient-specific number needed to harm

Suppose we are seeing an individual patient who we deem is specially susceptible to a therapeutic adverse reaction and we wish to foresee more precisely the risks about to be taken. This can be attained by applying the formula:

$$NNH/f_h \tag{16.11}$$

where NNH is the number needed to harm and f_h the fraction$_{harm}$.

NNH can be taken from literary sources [ondaparin study, for instance (*Section 16.1.4*)]. f_h can be taken from two sources:

- Empirical estimative

   One estimates that a patient like ours has a twofold risk of presenting the adverse reaction, relatively to a nontreated patient:

$$\frac{50}{2} = 25$$

- RRI (*Section 16.1.1.5*) from scientific literary sources consistent with our patient's case (ondaparin study, for instance):

$$\frac{50}{1.6} = 31$$

The above result means that it would be necessary to treat 25 (or 31) patients like ours, so that one of them would present an adverse reaction accountable to the treatment.

16.2.3 Patient-specific likelihood of being helped versus being harmed

Patient-specific LHH is an index that expresses the same information detailed in Section 16.1.5, adjusted for a specific patient. Calculated according to the ratio:

$$(1/\text{NNT} \times f_t):(1/\text{NNH} \times f_h) \tag{16.12}$$

$$(0.05 \times 2):(0.02 \times 1.6) \rightarrow 3:1$$

The above result means that ondaparin treatment has a probability of 3 of benefiting the specific patient, against a probability of 1 of harming him/her.

CHAPTER 17

Statistical assessment of diagnostic tests for the clinic

An investigator may perform a study aimed to test a new examination or diagnostic procedure. Similarly, a health professional may take the decision of which diagnostic test should be used for a specific patient, what contributions to expect from it, and how to interpret the available medical literature for supporting his/her conduct. The diagnostic tests can be performed with the aid of specific mathematical tools.

Suppose you see a patient at the emergency room (ER) with a complaint of precordial pain, started less than 6 hours ago. You ask for an electrocardiogram (EKG or ECG), inconclusive for acute coronary failure. Your institution has a nuclear medicine facility available 24/7, which is able to perform a SPECT (single-photon emission computed tomography) to document the perfusional deficit (if there is any). However, before ordering the test, you wish to gather more evidence that it might be useful for your diagnosis.

You search the literature and find a clinical trial consistent with your case: Myocardial perfusion scintigraphy (SPECT) in the evaluation of patients in the ER with precordial pain and normal or doubtful ischemic ECG. Study of 60 cases[1]. You follow the next step, which is to determine indexes of disease detection capacity and diagnostic significance of the considered test, based on the data of the trial, whose results are detailed in Table 17.1.

Based on these figures, the usefulness of the test can be determined through the indexes discussed in the following.

17.1 Detection capacity indexes

Detection capacity indexes determine a test capacity for detecting individuals with a condition and for NOT detecting those individuals who do NOT have it. These indexes are explained below.

---

[1] Bialostozky D., Lopez-Meneses M., Crespo L., *et al*, 1999. Myocardial perfusion scintigraphy (SPECT) in the evaluation of patients in the emergency room with precordial pain and normal or doubtful ischemic ECG. Study 60 cases. Arch Inst Cardiol Mex, 69(6), 534-45.

Practical Biostatistics
DOI: https://doi.org/10.1016/B978-0-323-90102-4.00008-4

Table 17.1 Results from single-photon emission computed tomography in a coronary failure trial [some data were changed for didactic purposes; (a), (b), (c) and (d) represent elements of some formulas detailed along the text].

	Patients with coronary failure	Patients without coronary failure	Total
Altered scintigraphy	19 (a)	10 (b)	29
Normal scintigraphy	6 (c)	25 (d)	31
Total	25	35	60

17.1.1 Sensitivity

Sensitivity determines the capacity of a test for detecting individuals with a condition, from a population in which all individuals have it. It is determined by the formula:

$$a/(a + c) \tag{17.1}$$

$$\frac{19}{(19 + 6)} = 0.76 = 76\%$$

The above result means that SPECT has a 76% probability of detecting coronary failure in an individual who actually has it. The larger this proportion, the smaller the probability of a false-negative yield and the greater the probability for condition non-existence in the case of a negative result.

17.1.2 Specificity

Specificity determines the capacity of a test in NOT detecting an altered test result in a population whose individuals do NOT have the considered condition. It is determined by the formula:

$$d/(b + d) \tag{17.2}$$

$$\frac{25}{(10 + 25)} = 0.71 = 71\%$$

The above result means that SPECT has a 71% probability of NOT detecting coronary failure in an individual who actually does NOT have it. By inference, it has a 29% probability of detecting this condition in an individual who does NOT have it.

The greater the sensitivity, the smaller the probability of a false-positive result and the greater the probability for condition existence in the case of a positive result.

17.1.3 Likelihood ratio

Likelihood ratio (LR) is an index which aggregates sensitivity and specificity, strengthening conclusions inferable from both. It can be expressed in two ways as detailed below.

17.1.3.1 Positive likelihood ratio

Positive LR (LR +) determines the probability of a test in detecting condition A instead of condition B, the former presenting the capacity of changing test results as well. It is determined by the formula:

$$\text{sensitivity}/(100 - \text{specificity}) \tag{17.3}$$

$$\frac{76\%}{(100\% - 71\%)} = 2.6$$

The above result means that SPECT has a 2.6 greater probability of detecting coronary failure than other causes potentially associated with an altered SPECT result such as myocarditis.

As a general rule, the following ranges for LR interpretation can be adopted:
- $>10 \rightarrow$ the test has a high capacity of detecting the suspected condition;
- $\sim1 \rightarrow$ the test has a limited capacity of detecting the suspected condition;
- $<0,1 \rightarrow$ the test has a low capacity of detecting the suspected condition.

17.1.3.2 Negative likelihood ratio

Negative LR (LR -) determines the probability of a test in detecting condition B, which could also change test results, instead of condition A. It is determined by the formula:

$$(100 - \text{sensitivity})/\text{specificity} \tag{17.4}$$

$$\frac{(100 - 76)}{71} = 0.33 \, (or \, 33\%)$$

The above result means that SPECT has a 33% probability of detecting a condition other than coronary failure, rather than coronary failure itself.

According to these detection capacity indexes, SPECT has a good capacity for coronary failure detection in a patient with a less than 6 hours precordial pain and a doubtful EKG.

17.2 Diagnostic significance indexes

Diagnostic significance indexes determine the capacity of a positive test result in representing a condition existence and of a negative yield in not representing it. These indexes are as follows:

17.2.1 Pretest probability (prevalence)

Pretest probability determines the proportion of individuals with a condition, relatively to a population under risk. It is determined by the formula:

$$(a + c)/(a + b + c + d) \qquad (17.5)$$

$$\frac{25}{60} = 0.41 = 41\%$$

Pretest probability usefulness is taking part in pretest odds calculation.

17.2.2 Pretest odds

Pretest odds determine the odds of an individual who belongs to a population under risk, in presenting a certain condition. The higher the prevalence, the higher the odds. It is determined by the formula:

$$\text{prevalence}/(1 - \text{prevalence}) \qquad (17.6)$$

$$\frac{0.41}{0.59} = 0.69$$

Results range from 0 (null odds) to 1.0 (the highest possible odds). Pretest odds usefulness is taking part in posttest odds calculation.

17.2.3 Posttest odds

Posttest odds determine the odds of an individual who presents a positive test result, in actually having the suspected condition. It is determined by the formula:

$$\text{pretest odds} \times \text{positive likelihood ratio} \qquad (17.7)$$

$$0.69 \times 2.6 = 1.8$$

The above result represents a 1.8:1 posttest odds for the actual presence of coronary failure in an individual presenting a positive SPECT.

17.2.4 Posttest probability

Posttest probability determines the proportion of individuals presenting a positive test result, which represents subjects who actually have the suspected condition. It is determined by the formula:

$$\text{posttest odds}/(\text{posttest odds} + 1) \tag{17.8}$$

$$\frac{1.8}{2.8} = 0.64 = 64\%$$

Posttest probability expresses essentially the same as posttest odds, but from a collective perspective.

17.2.5 Positive predictive value

Positive predictive value determines the proportion of individuals presenting a positive test result, which represents subjects who actually have the suspected condition. It is determined by the formula:

$$a/(a + b) \tag{17.9}$$

$$\frac{19}{29} = 0.65 = 65\%$$

Positive predictive value expresses essentially the same as posttest probability, through a different mathematical approach.

17.2.6 Negative predictive value

Negative predictive value determines the proportion of individuals presenting a negative test result, which represents subjects who do NOT have the suspected condition. It is determined by the formula:

$$d/(c + d) \tag{17.10}$$

$$\frac{25}{31} = 0.80 = 80\%$$

It expresses the probability of a test in NOT yielding a false-positive result.

Based on the overall above results, one can infer that a positive myocardial perfusion scintigraphy (SPECT) test points to a good probability for coronary failure in a patient with precordial pain complaint lasting less than 6 hours and an EKG inconclusive for ischemic heart disease.

The determination of the prevalence of a condition is essential for the calculation of posttest odds/posttest probability indexes. An available source from where this information could be extracted was used in the above examples (a clinical trial involving the investigated procedure). However, different sources can be explored if there are no immediate consistent studies available, as follows:

For example, your hospital records show that 34 patients are diagnosed with coronary failure out of every 100 patients admitted through the ER.

- Personal or institutional databank
- Statistical database from public health institutions;
- Scientific studies of prevalence determination for several different conditions.

However, presuming the gathering of figures from several different sources will be a likely event, it would be advisable to ask for the support of a biostatistician.

Examples of detection capacity and diagnostic significance indexes already published:

Serum ferritin determination for iron deficiency anemia diagnosis: (1) sensitivity, 90.4%; (2) specificity, 84.7%; (3) positive predictive value, 73%; (4) negative predictive value, 95%; and (5) posttest odds, 2.6.

D-dimer serum levels > 1092 ng/mL for deep venous thrombosis detection, in patients with stroke sequelae: LR, 3.1.

Manual device determined serum heart-specific troponin T levels, for myocardial infarction detection within 2 hours of clinical onset: (1) LR +, 6.3 and (2) LR −, 0.8.

CHAPTER 18

Systematic reviews and meta-analyses

18.1 Systematic review

Systematic review is a comprehensive, protocol-based, and reproductible bibliographic review of randomized clinical trials (primary studies). It aims to test a clearly stated investigator's hypothesis and is a preceding stage to meta-analysis. Systematic reviews can be performed in following three stages:

18.1.1 Problem formulation

Investigator's hypothesis (see Chapter 1: Investigator's Hypothesis and Expression of Its Corresponding Outcome) must be explicitly formulated, as for any other scientific study.

18.1.2 Primary studies search and selection

Performed by a member of the research team named "searcher." Following are the proposed guidelines:

- Search criteria and determination of minimum quality level of primary studies must be stated in the study protocol: (1) primary studies must be randomized, (2) studied object of the primary study must have been compared to an adequate control, (3) studied sample in the primary study must be representative of the target-population focused by the systematic review investigator, (4) adjustments for eventual differences among the several primary studies elements must be considered, and (5) primary study subjects must be sufficiently homogeneous regarding prognosis.
- Primary studies must be cross-consistent regarding the following aspects: (1) population type, (2) type of condition, (3) type of intervention, (4) methodology, and (5) type of outcome.
- Search results must be reproduced by an independent reviewer and this reproduction must be statistically validated.
- Language restrictions must be avoided.
- Nonnormal distribution studies (see Chapter 13: Determination of Normality or Nonnormality of Data Distribution) must be ruled out.
- Search must be thorough and every possible source must be considered: databases, sponsoring agencies, pharmaceutical companies studies, regulatory agencies, clinical trials registries, and the so-called "gray literature."

Practical Biostatistics
DOI: https://doi.org/10.1016/B978-0-323-90102-4.00003-5

Discrepancies between searcher and independent reviewer findings can be solved through statistical methodology or a second independent reviewer.

18.1.3 Primary studies data extraction

The searcher's objective in this stage is to organize extracted data from primary studies to determine if meta-analysis is feasible. Its methodology is based on the following guidelines:

- Analysis must be performed according to a specially designed form.
- Search results must be reproduced by an independent reviewer.
- Extraction should be "blind," that is, the name of the authors of primary studies must be concealed.
- Data extraction process can be extended through direct contact with the authors of primary studies (this procedure would obviously nullify the "blindness" status of the research and this should be stated).
- Extracted data must be tabulated for insertion in a suitable software.

Discrepancies between searcher and independent reviewer findings can be solved through statistical methodology or a second independent reviewer.

18.2 Meta-analysis

Meta-analysis is a statistical procedure that aims to determine efficacy or nonefficacy, generally of a medical intervention, through analysis of data across primary studies selected by systematic review, to generate an average pooled estimate of their effect sizes (pooled estimate of effects—see further). In principle, meta-analyses afford a more powerful statistical testing than respective primary studies separately. The following endpoints can be evaluated through meta-analyses: (1) relative risk, odds ratio, absolute risk reduction, number needed to treat; (2) scores; (3) sensitivity and specificity; and (4) P values.

A meta-analysis is not a mere arithmetic mean of the results from the different primary studies, but it is an elaborate methodology that aims to determine the weighted mean of effect sizes of pooled primary studies, selected by the means of a systematic review. Meta-analysis can be performed in following four stages:

18.2.1 Publication bias detection

Publication bias can be investigated with the aid of a funnel plot graph (Fig. 18.1).

The weighted mean of effect sizes (pooled estimate of effects) corresponds to funnel axis. The larger n of a primary study, the larger its corresponding weight in a pooled estimate of effects determination. This is the reason their corresponding blocks (*see further*) tend to stand closer to the funnel axis. The opposite is true for primary studies with a smaller n, due to their larger variability of results.

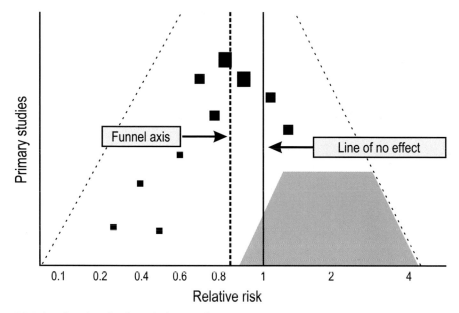

Figure 18.1 Landmarks of a funnel plot graph.

This type of distribution tends to generate a characteristic funnel-shaped plotting. Publication bias can be visualized through graphic paucity of primary studies (trapezoid field), which are generally associated with negative results and/or a small *n*. As such, the closer to a pyramid shape the funnel plot is, the weaker the publication bias. Publication bias level and the extent to which it compromises meta-analysis quality must be determined by the investigators.

18.2.2 Heterogeneity analysis

Primary studies heterogeneity caused by between-study differences is an expected circumstance. Its analysis is crucial for defining whether selected primary studies pooling is fit for meta-analysis. Heterogeneity can manifest in two ways, with corresponding rectifying approaches:

- Clinical heterogeneity

    It requires assessment based on clinical grounds.

- Methodological heterogeneity

    It requires statistical quantification. In this case, the investigators must assume heterogeneity among primary studies themselves as the null hypothesis (H_0). Therefore rejecting H_0 means that there would be sufficient homogeneity among these studies ($P < .10$ is generally acceptable). The limitation of statistical tests is that their power to detect statistically significant homogeneity is weakened among primary studies with a small *n* and with pools of few primary studies.

If homogeneity is not demonstrated, it is not advisable to upgrade the systematic review to meta-analysis.

18.2.3 Summarized statistical determination

Summarized statistical determination corresponds to effect size specification of each individual primary study. It is necessary step for meta-analysis weighted mean determination.

18.2.4 Meta-analysis performance and expression

Pooled estimate of individual effect sizes of primary studies can be calculated as a weighted mean:

$$\text{meta-analysis weighted mean} = \Sigma T_i W_i / \Sigma W_i \qquad (18.1)$$

i=individual, T_i= effect size attributed to i primary study, W_i= weight attributed to i primary study.

Standard error of the mean (see Chapter 12: Measures for Results Expression of a Clinical Trial) of this pooled estimate can be used to determine 95% confidence interval (see Chapter 15: Correlating Sample Data with the General Population—95% Confidence Interval) and corresponding P.

One of the most popular ways of graphically representing meta-analysis results is the *forest plot* (confidence intervals plotting). Its graphical elements are depicted in Fig. 18.2.

- Block

    The block represents the weighted mean (point estimate) of i primary study x. Its relative size and proximity to the meta-analysis pooled estimate (meta-analysis weighted mean) are proportional to i primary study x relative weight.

- 95% Confidence interval line

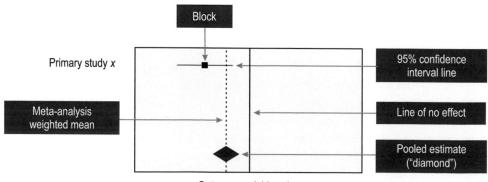

Figure 18.2 Landmarks of a forest plot graph.

95% Confidence interval line represents *i* primary study confidence interval. The width of the former is expected to be inversely proportional, both to *i* primary study's weight and proximity to meta-analysis pooled estimate. To be significant in the context of a meta-analysis, it must not touch the "line of no effect."

- Significance line (line of no effect)

  Significance line represents neutrality of the outcome variable value.

- Meta-analysis weighted mean

  Meta-analysis weighted mean vertically represents the weighted mean of effect sizes, obtained by the meta-analysis.

- Pooled estimate

  Pooled estimate represents the weighted mean of effect sizes, obtained by the meta-analysis. Its width corresponds to its 95% confidence interval.

An example of a hypothetical meta-analysis on the efficacy of a vaccine, based on the analysis of the relative risk of being protected by the vaccine supposed to prevent the disease, is shown in Fig. 18.3.

A pooled estimate demonstrates a protective effect provided by the vaccine. Nevertheless, the statistical significance of this result, *P*, must also be determined. Additional data to the forest plot are (1) *n*, (2) mean and standard deviation, (3) *P* of each primary study, (4) meta-analysis overall 95% confidence interval, and (5) heterogeneity test result (with statistical significance level).

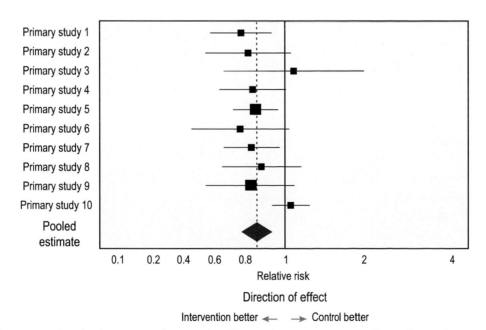

Figure 18.3 Result of a meta-analysis on the efficacy of a vaccine, expressed as a forest plot.

18.3 Options if meta-analysis performance is not possible

- Keep the research as systematic review only
- Perform a subgroup meta-analysis

    In a subgroup meta-analysis, a heterogeneous population of primary studies is subdivided into two homogeneous subgroups. The result can be expressed as a forest plot graph. A meta-analysis involving 10 primary studies considered as heterogeneous is exemplified in Fig. 18.4. From this first meta-analysis, two separate pooled estimates are obtainable by dividing the different primary studies in subgroups (Fig. 18.5).

- Perform sensitivity analysis

    In a meta-analysis context, sensitivity analysis consists of a reanalysis of the data set of primary studies, aiming to determine if identifying possible confounders (type of intervention, subject profile, outcome variables) or decreasing heterogeneity among them could lead to a different final outcome or interpretation. This procedure might allow a new meta-analysis trial.

- Perform meta-regression analysis

    Meta-regression analysis is a statistical procedure that aims to identify and quantify sources of heterogeneity among primary studies. Tools such as multiple linear regression (see Chapter 19: Correlation and Regression) or logistic regression are used to explore the relationship among the parameters of primary studies, such as geographical location, subjects' age, and effect size.

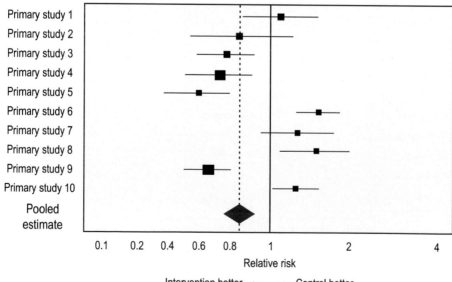

Figure 18.4 Result of a meta-analysis on 10 heterogeneous primary studies.

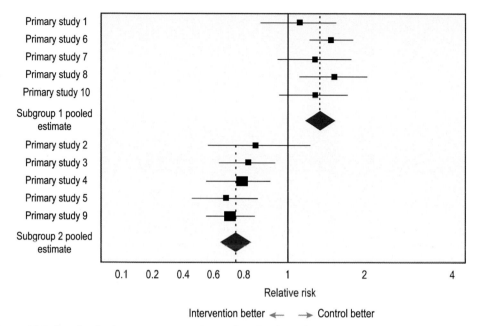

Figure 18.5 Result of subgroup meta-analysis, after division of primary studies as subgroups.

18.4 Systematic review/meta-analysis limitations

- Publication bias: (1) citation bias (positive studies are more likely to be published), (2) language bias (primary studies are more likely to be published in English), (3) so-called "gray literature" papers are harder to find, (4) positive studies are published more swiftly than negative studies, (5) larger n studies are more likely to be published, and (6) smaller n studies are more likely to be published if they present positive results.
- Searcher and independent reviewer learning curve

  Inexperienced searchers and independent reviewers tend to make assessment and methodological errors during their research. Participation of an experienced researcher might compensate for this limitation.

18.5 Summary of systematic review/meta-analysis stages

- Systematic review: (1) problem formulation, (2) primary studies search and selection, and (3) primary studies data extraction.
- Meta-analysis: (1) publication bias detection, (2) heterogeneity analysis, (3) summarized statistics determination, and (4) meta-analysis performance and expression.
- Options if meta-analysis performance is not advisable: (1) keep the research as systematic review only, (2) perform subgroup meta-analysis, (3) perform sensitivity analysis, and (4) perform meta-regression analysis.

CHAPTER 19

Correlation and regression

Sometimes, the investigator's hypothesis does not involve searching for statistically significant differences between groups, but whether two different study variables relate to each other. And, if they significantly do, then predicting the value of one variable based on the value of the other one might be feasible. Two mathematical tools are available to accomplish these objectives: correlation and regression.

Note: determination of the parameters presented in this chapter can be performed through widely available online resources. We will detail corresponding calculations in a conventional manner, for didactical purposes.

19.1 Correlation

Correlation has two goals: (1) to quantify the degree of connection between a pair of variables and (2) to determine the direction of this relationship (see further). Biological plausibility demands that, in the clinical setting, the "strongest" variable influences the "more susceptible" one. This proposition implies two opposing concepts:

* Independent variable

    The independent variable is the one that influences the dependent variable, but in turn, is not influenced by the latter. In didactic terms, it is the "dominant" variable. For example, body temperature (independent variable) influences heart rate (dependent variable), rather than the opposite. By convention, it is identified as x and graphically represented by the horizontal axis (x-axis).

* Dependent variable

    The dependent variable is the one that is influenced by the independent variable, but in turn, does not influence the latter. In didactic terms, it is the "submissive" variable. For example, heart rate (dependent variable) is influenced by body temperature (independent variable), rather than the opposite. By convention, it is identified as y and graphically represented by the vertical axis (y-axis).

For example, body weight and mean arterial blood pressure data from a cohort of 26 overweight middle-aged men are tabulated in Table 19.1. Biological plausibility implies that body weight is the independent variable (x) and mean arterial blood pressure is the dependent variable (y).

Firstly we must check if there is a hint of a relationship between both variables through a graph and quantify it (Fig. 19.1).

Practical Biostatistics
DOI: https://doi.org/10.1016/B978-0-323-90102-4.00018-7

Table 19.1 Data from a cohort of 26 overweight middle-aged men.

Patient	Body weight (kg)[a] (x)	Mean arterial blood pressure (mmHg) (y)
1	20	80
2	30	78
3	40	90
4	50	92
5	60	76
6	70	78
7	80	86
8	90	76
9	100	108
10	110	74
11	120	85
12	130	108
13	140	110
14	150	88
15	160	90
16	170	80
17	180	118
18	20	150
19	30	89
20	40	90
21	50	75
22	60	78
23	70	108
24	80	145
25	90	198
26	100	149
Mean	142.8	99.9

[a]Figures are unrealistic and meant to only demonstrate the proposed concepts in a didactic fashion.

Based on a simple visual analysis, it is possible to notice a correlation between body weight and mean arterial blood pressure variables, that is, the higher the body weight, the higher mean arterial blood pressure. To better quantify this relationship, an index called *correlation coefficient* must be determined. One of the most commonly used is *Pearson's correlation coefficient* (*r*). This coefficient is determined by the formula:

$$r = \Sigma(x - \bar{x})(y - \bar{y})/\sqrt{\Sigma(x-\bar{x})^2 \Sigma(y-\bar{y})^2} \tag{19.1}$$

Where $\bar{x} = x$ values mean, and $\bar{y} = y$ values mean.

$$r = \frac{(20 - 142.8) + (30 - 142.8) + \ldots + (198 - 99.9) + (149 - 99.9)}{\sqrt{(20-142.8)^2 + (30-142.8)^2 + \ldots + (198-99.9)^2 + (149-99.9)^2}} = +0.57$$

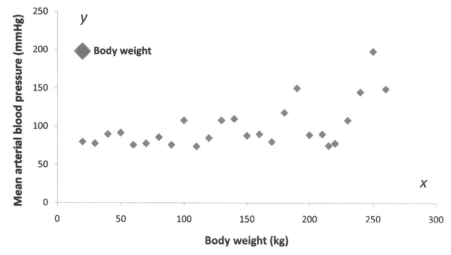

Figure 19.1 Scatterplot based on the data from Table 19.1.

r ranges from −1 to +1 (− denotes a negative *direction* and + a positive *direction*). Possible inferences based on these data are as follows:

- The closer to 0, the weaker the correlation (dependent variable is "indifferent" to independent variable changes).
- The closer to 1 (positive or negative), the stronger the correlation; the dependent variable is expected to change as much as the independent variable does, in the following manner:
  - the closer to −1, the more the dependent variable distances from the independent variable (if the latter increases, then the former decreases and vice versa);
  - the closer to +1, the more both variables change in parallel (if the independent variable either increases or reduces, the dependent variable follows it).

Empirically, the degree of correlation can be classified as (either as negative or positive direction):

- 0−0.03: null
- 0.04−0.35: weak
- 0.36−0.65: moderate
- 0.66−0.95: strong
- 0.96−1.0: very strong

In the above example, $r = +0.57$. The numerical value itself represents a moderate correlation between body weight and mean arterial blood pressure. Positive direction on its turn implies two possibilities: (1) mean arterial blood pressure is supposed to increase as body weight does so and (2) mean arterial blood pressure is supposed to decrease as body weight does so.

19.2 Regression

Regression[1] (linear regression) takes correlation one step further, by actually predicting the value of a dependent variable based on the value of an independent variable, as it quantifies the strength and direction of this prediction. This method is based on the so-called *regression line*, determined by the *linear regression formula*:

$$y' = 4 + 2x \qquad (19.2)$$

Where y' = value of the dependent variable to be predicted, and x = independent variable.

That means, if $x = 2$, then $y' = 8$. Graphically, this line would be represented as in Fig. 19.2:

The regression line crosses as closely as possible all the intersection points of a graph. By doing so, it is expected to represent the trend of pooled data, as well as its direction (ascending, flat, or descending line). Two important elements of the regression line are as follows:

- Slope

    The slope represents the steepness of the line. It informs on the size of the influence x has over y, that is, the steeper the line, the greater this influence.

- y-intercept

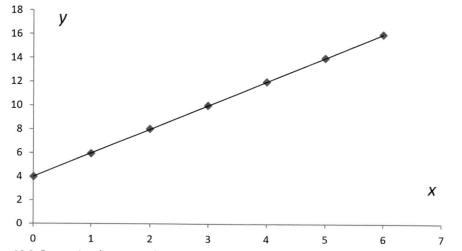

Figure 19.2 Regression line, traced according to the intersection of independent and dependent variables. It is presumed that these variables present a linear relationship.

[1] The term "regression" does not imply a temporal dimension to the problem; it only represents a historical aspect of the development of this tool.

The y-intercept is the point where regression line touches y-axis. It informs on the value of the dependent variable when the independent variable equals 0.

To predict a dependent variable it is first necessary to build a linear regression graph by adding a regression line to the correlation scatterplot of the studied population. Let us use the example from Fig. 19.1 (Fig. 19.3).

Interpretation of a linear regression graph might depend upon visual analysis (the closer to the intersection points, the greater the predictability) and biological plausibility. One can also use the linear regression formula to predict mean arterial blood pressure (y) based on any value of body weight (x) by replacing 4 for a, representing the y-intercept and 2 by b or $-b$, representing the slope:

$$y' = a + bx \qquad (19.3)$$

or

$$y' = a + (-bx) \qquad (19.4)$$

Where y' = value to be predicted, a = y-intercept, b = ascending slope (positive direction), $-b$ = descending slope (negative direction), and x = independent variable.

a and b are in fact *regression coefficients*, which must be calculated before finding y'. The formulas for their calculations are, respectively:

$$a = y' - bx \qquad (19.5)$$

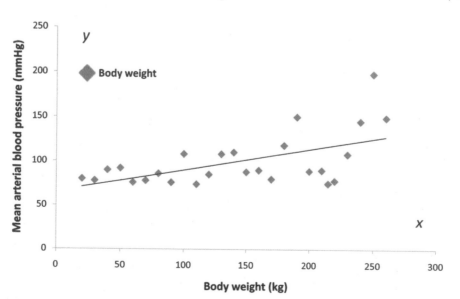

Figure 19.3 Regression line traced on the scatterplot from Fig. 19.1.

$$b = \Sigma(x - \bar{x})(y - \bar{y})/\Sigma(x - \bar{x})^2 \tag{19.6}$$

Where $\bar{x}$ = x values mean, and $\bar{y}$ = y values mean.

By applying the data from Table 19.1, we have the following values:

$$a = 99.9 - (0.235 \times 142.8) = 66.2$$

$$b = \frac{(20 - 142.8)(80 - 99.9) + \ldots + (100 - 142.8)(149 - 99.9)}{(20 - 142.8)^2 + (30 - 142.8)^2 + \ldots + (90 - 142.8)^2 + (100 - 142.8)^2} = 0.235$$

Suppose we wish to predict mean arterial blood pressure for subject 12 from our cohort. His weight is 130 kg:

$$y' = a + bx$$

$$y' = 66.2 + 0.235(130) = 96.7 \text{mmHg}$$

Nevertheless, it is also necessary to determine how much this mean arterial blood pressure value (y') is explainable by his body weight (x), according to this model. This is achievable through *coefficient of determination* (R^2):

$$R^{2*} = r^2 \tag{19.7}$$

\Where r = Pearson's correlation coefficient. *Read as "r two."

$$R^2 = +0.57^2 = 0.32 \text{ or } 32\%$$

This figure means that, from all possible reasons for patient 12 to present a 96.7 mmHg mean arterial blood pressure (y'), 32% can be explained by his 130 kg body weight (x), according to linear regression model.

19.3 Multiple linear regression

Eventually, two or more independent variables are available for predicting the dependent variable. These independent variables can be considered in combination for the dependent variable estimate, through *multivariable analysis* (*Chapter 8: Increasing Accuracy in Observational Studies*). There are plenty of available tools for performing multivariable analysis and we will focus on *multiple linear regression* to demonstrate this type of resource. Let us detail multiple linear regression by extending the example used in Table 19.1 (Table 19.2).

Table 19.2 Data from the cohort of 26 overweight middle-aged men, considering diastolic blood pressure as a second independent variable.

Patient	Body weight (kg) (x_1)	Diastolic blood pressure (mmHg) (x_2)	Mean arterial blood pressure (mmHg) (y)
1	20	90	80
2	30	82	78
3	40	99	90
4	50	99	92
5	60	80	76
6	70	82	78
7	80	95	86
8	90	80	76
9	100	110	108
10	110	80	74
11	120	89	85
12	130	115	108
13	140	120	110
14	150	100	88
15	160	100	90
16	170	90	80
17	180	129	118
18	20	160	150
19	30	99	89
20	40	100	90
21	50	85	75
22	60	88	78
23	70	118	108
24	80	153	145
25	90	202	198
26	100	180	149
Mean	142.8	108.6	99.9

By scatterplotting these data altogether and adding a trendline for each independent variable, we can verify that x_1 and x_2 have different effects on y, as represented in Fig. 19.4.

Notice how the second independent variable—diastolic blood pressure (x_2) —generates by itself a steeper slope toward greater mean arterial blood pressure values. Hence, if we wish to predict mean arterial blood pressure in a more "realistic" way, it would be advisable to consider both independent variables combined. If the investigator decides to do so, he/she will have two options, according to the presence or absence of interaction between (or among) independent variables:

• Independent variables do not interact

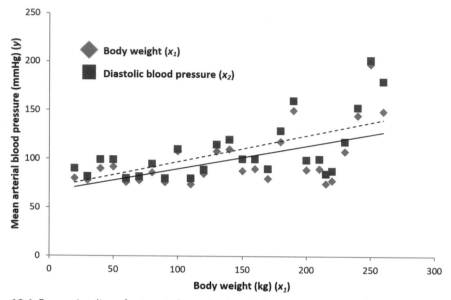

Figure 19.4 Regression lines for two independent variables: x_1 and x_2, traced on a scatterplot built for Table 19.2 (continuous line represents x_1 and dashed line x_2).

In this case, the *multiple linear regression formula* can be directly applied. With this resource, the dependent variable tends to change *linearly* (i.e., proportionally) with the weighted sum of independent variables. In multiple linear regression formula, *partial regression coefficients*—b_1, b_2, ... b_k—are used, instead of regression coefficients:

$$y' = a + b_1x_1 + b_2x_2 + \ldots + b_kx_k \tag{19.8}$$

Where y' = value to be predicted (dependent variable), a = y intercept, b_1 = partial regression coefficient to x_1, x_1 = independent variable 1, b_2 = partial regression coefficient to x_2, x_2 = independent variable 2, b_k = partial regression coefficient to x_k, x_k = a given independent variable.

In this setting, changes in one independent variable, for example, x_1, change y' on its own and not through indirect influence on other independent variables. In medical sciences, this situation represents a minority of cases and the current example is no exception.

• Independent variables interact

In the above example, body weight and diastolic blood pressure—our so-called "independent" variables—are in fact mutually influenced. It is known that patients with an elevated body weight are prone to present a higher diastolic blood pressure. Therefore body weight would be an "independent variable" to diastolic blood pressure, now the "dependent variable." In this scenario, the application of multiple linear regression formula cannot be directly performed. Partial regression coefficients must be mathematically adjusted according to the degree of reciprocal influence strength among the so-called independent variables that the former ones represent (method detailed elsewhere). This strength degree must be statistically validated, before including adjusted partial regression coefficients in the analysis. In medical sciences, this situation represents the majority of cases.

CHAPTER 20

Per-protocol analysis and intention-to-treat analysis

A research team is expected to strictly adhere to a randomized study protocol. However, even though the research subject may have been fully oriented, occasionally he/she can fail to comply with the instructions along the trial. In fact, investigators have poor control over this phenomenon because the research subject, once leaving medical visit and returning to daily routine, will be as subject to failures as in any average medical treatment (forgetfulness, self-neglect, nonadherence to study visits).

This situation generates a significant dilemma for the research team, which will be forced to choose one of the following approaches for the final analysis of study results, both with their advantages and drawbacks:

20.1 Per-protocol analysis

In per-protocol analysis, investigators are absolutely strict regarding a subject's adherence to study protocol:

- Advantage

    Reliability of the study will be better preserved since errors generated by research subjects will be avoided. Therefore all subjects who fail to strictly comply to study protocol are expected to be excluded from the final analysis.

- Limitations

    Total adherence to a prescribed therapy or medical orientation is not frequently seen in daily practice, but small flaws which, in general, do not preclude completion of the treatment or its success. If the research team rejects this pattern, it would be alienating itself from so-called "real world" situations.

    In randomized studies, if it is established that only fully adherent subjects should be considered for final analysis, a selection criterion would be being generated, which could corrupt randomness. This situation would, in itself, correspond to a bias.

    If only fully adherent subjects are considered for final analysis, the study n could be severely diminished.

Practical Biostatistics
DOI: https://doi.org/10.1016/B978-0-323-90102-4.00021-7

20.2 Intention-to-treat analysis

In intention-to-treat analysis, investigators assume a controlled tolerance attitude regarding subject's adherence flaws to study protocol:

- Advantages

    Tolerating subjects' adherence flaws to study protocol—provided they are NOT severe flaws—may afford a more reliable forecast regarding "real world" practice, where the tested drug or vaccine will be actually used. Nevertheless, it is expected that randomization will itself evenly dilute these flaws, hence minimizing random error.

    Randomization protection by maintaining research subjects in the study groups they were initially allocated to, during final analysis.

    A partially adherent subject may have behaved so due to an adverse event related to the tested drug or vaccine. By keeping him/her in the final analysis, this important information would be preserved, thus avoiding safety analysis error.

    Protocol nonadherent subjects may represent individuals who are also nonadherent in other aspects regarding their own health. Thus by keeping them in the final analysis, the generation of artifactual "sicker" or "healthier" samples, with accompanying bias, would be avoided.

- Limitation

    The inclusion of subjects not fully adhered to study protocol in the final analysis would move the conclusion of the study further away from the truth.

It is important to establish the limit between a protocol flaw and nonadherence to study protocol (which should be characterized as a dropout). This parameter must be stated by the research team in the study protocol, based on the following suggested parameters: (1) tolerable frequency of posology flaws, (2) tolerable frequency of nonattendance to study visits, and (3) research subjects should not miss the visits scheduled for measurements of efficacy and safety endpoints.

Intention-to-treat is, therefore, a trial results analysis type in which partial protocol adherence is tolerated, in exchange for its respective advantages. In per-protocol analysis, only subjects strictly adhered to study protocol are taken into account for final analysis.

Both intention-to-treat and per-protocol analyses can be performed in the same study. For example, a new nucleoside reverse transcriptase inhibitor efficacy in elevating CD4 + T cell blood count in HIV + patients (group A) is tested, against a reference inhibitor belonging to the same class (group B). Results are detailed in Table 20.1.

Table 20.1 Results of CD4 + T cell counts (cells/mm^3) for groups A and B, as per-protocol and intention-to-treat analyses.

	Per-protocol analysis		Intention-to-treat analysis	
n	30	39	44	45
Group A	503	702	439	490
Group B	550	608	490	600

One possible solution for the dilemma of which approach to adopt is using both approaches. If results coincide, study conclusion will be strengthened. In the case they do not, then intention-to-treat analysis might be preferable, for two reasons: (1) better-preserved randomization, and (2) in general, a Type II error is safer for Medicine than a Type I error.

Bibliography

Suggested reading (Part 5)

Bialostozky, D., Lopez-Meneses, M., Crespo, L., et al., 1999. Myocardial perfusion scintigraphy (SPECT) in the evaluation of patients in the emergency room with precordial pain and normal or doubtful ischemic ECG. Study 60 cases. Arch. Inst. Cardiol. Mex. 69 (6), 534—545.

Canadian Medical Association Journal, 2017. Basic statistics for clinicians. www.cmaj.ca. (Accessed 4 May 2017).

Centre for Evidence Based Medicine, 2016. Statistics for laboratory medicine. Department of Medicine, Toronto General Hospital. www.cebm.utoronto.ca (Accessed 3 May 2016).

Center for Health Evidence. Systematic reviews and meta-analyses. www.cche.net/default.asp.

Drummond, J.P., Silva, E., 2012. Medicina Baseada em Evidências — Novo Paradigma Assistencial e Pedagógico. Atheneu, Rio de Janeiro (Chapter 4).

Estrela, C., 2001. Metodologia Científica - Ensino e Pesquisa em Odontologia, first ed. Editora Artes Médicas, Porto Alegre (Chapter 7).

Everitt, B.S., et al., 2005. Encyclopaedic Companion to Medical Statistics. Hodder Arnold, London.

Everitt, B., 2006. Medical Statistics From A to Z. A Guide for Clinicians and Medical Students, second ed. Cambridge University Press, London.

Everitt, B.S., 2008. The Cambridge Dictionary of Statistics in the Medical Sciences. Cambridge, London.

Green, S., 2005. Systematic reviews and meta analysis. Singap. Med. J. 46 (6), 270—274.

Hulley, S.B., et al., 2001. Designing Clinical Research: An Epidemiological Approach, second ed. Lippincott Williams & Wilkins, Philadelphia, PA.

Jaeschke, R., Guyatt, G., Shannon, H., et al., 2011. Assessing the effects of treatment: measures of association. Can. Med. Assoc. J. 152, 351—357.

Laupacis, A., Sackett, D.L., Roberts, R.S., 1988. An assessment of clinically useful measures of the consequences of treatment. N. Engl. J. Med. 318 (26), 1728—1733.

Merriam-Webster Online Dictionary. www.merriam-webster.com.

Mitchell, H., Katz, M.H., 1999. Multivariable Analysis. A Practical Guide for Clinicians. Cambridge University Press, London.

Montori, V.M., Swiontkowski, M.F., Cook, D.J., 2003. Methodologic issues in systematic reviews and meta-analysis. Clin. Orthop. Rel. Res. (413), 43—54.

Oliveira, G.G., 2006. Ensaios Clínicos. Princípios e Prática. Editora Anvisa, Brasilia.

Pai, M., McCulloch, M., Gorman, J.D., et al., 2004. Systematic reviews and meta-analysis: an illustrated, step-by-step guide. Natl. Med. J. India 17, 86—95.

Petrie, A., Bulman, J.S., Osborn, J.F., 2002. Further statistics in dentistry. Part 6: Multiple linear regression. Br. Dent. J. 193 (12), 675—682.

Sackett, D.L., et al., 2001. Evidence-Based Medicine: How to Practice and Teach EBM, second ed. Elsevier Health Sciences, Amsterdam.

Sauerland, S., Seiler, C.M., 2005. Role of systematic reviews and meta-analysis in evidence-based medicine. World J. Surg. 29, 582—587.

Schmuller, J., 2009. Statistical Analysis With Excel for Dummies, second ed. Wiley Publishing, Hoboken, NJ.

The Cochrane Collaboration. Cochrane Handbook for Systematic Reviews of Interventions (5.0.1).

Appendix: Overview of study types for human health investigation

The classification detailed below is not a definitive one, for not enlisted study types or mixed study types can be implemented depending upon the investigator's objectives and resources.

1. Collective health
   a. Disease frequency measures
      i. Simple count
      ii. Prevalence: (1) point prevalence and (2) period prevalence
      iii. Incidence: (1) cumulative incidence and (2) incidence rate
   b. Health indicators
      i. Survival
      ii. Mortality
         • Crude mortality rate
         • Specific mortality rate
         • Mortality proportion
         • Lethality rate
         • Mortality indicators according to cause of death: (1) mortality proportion due to cause of death and (2) mortality rate due to cause of death
      iii. Life indicators
         • Life expectancy: (1) simple survival and (2) person-year survival rate and person-year death rate
         • Years of potential life lost
      iv. Morbidity indicators: (1) incidence and (2) prevalence
   c. Epidemiological studies: (1) ecological studies, (2) cross-sectional studies, and (3) longitudinal studies
   d. Pharmacoeconomics
      i. Cost-oriented timing: (1) costs standardization and (2) discount
      ii. Costs minimization analysis
      iii. Cost−efficacy analysis
         • Cost−consequence analysis
         • Cost−efficacy ratio: (1) simple cost−efficacy ratio, (2) cost−efficacy ratio for percentual unit of success, (3) cost−efficacy ratio for percentual of an additional success, and (4) cost−efficacy increment ratio
      iv. Utility
         • Utility analysis: (1) scaling, (2) standardized game, and (3) time negotiation
         • Cost−utility analysis: (1) utility gain and (2) cost−utility ratio

  v. Financial resources
    - Cost—benefit analysis: (1) liquid benefit and liquid cost and (2) cost——benefit and benefit—cost ratios
    - Human capital
    - Return rate
2. Observational studies
  a. Case—control studies: (1) odds ratio and (2) number needed to harm
  b. Cohort studies: (1) relative risk and (2) number needed to harm
  c. Retrospective cohort studies
3. Intervention studies
  a. Reference standard: (1) noncomparative studies and (2) comparative studies
  b. Relation between samples and of a sample with itself:
    i. Nonpaired sample studies
    ii. Paired sample studies: (1) self-pairing, (2) natural pairing, and (3) artificial pairing
  c. Awareness of tested drug, vaccine, or exam: (1) open studies, (2) single-blinded studies, and (3) double-blinded studies
  d. Study subjects allocation method: (1) nonrandomized studies and (2) randomized studies
  e. Follow-up method: (1) parallel studies and (2) crossover studies
  f. Subgroup analysis
4. Systematic reviews and meta-analyses
5. Correlation and regression

Glossary

Absolute risk Possibility that a disease-free subject submitted to a known exposure factor might present a given condition in a certain timespan.

α Statistical significance level corresponding to the highest tolerable cutoff for type I error (generally 0.05).

Alternative hypothesis Hypothesis against which an antagonistic null hypothesis (H_0) (see *null hypothesis*) is tested. Alternative hypothesis (H_1) normally prevails if H_0 is rejected.

Analytical studies Study type superfamily whose aim is to establish correlation strength between a condition and a factor putatively associated with its origin and/or natural history. Observational and intervention studies are their main study types.

β Statistical significance level corresponding to the highest tolerable cutoff for type II error (generally 0.20).

Bias Deviation of results and inferences from truth, due to systematic error (see *systematic error*).

Chance The possibility of a particular outcome in an uncertain situation (lit.).

Case A term used to refer to an individual in a population who has the condition of interest.

Central limit theorem Mathematical principle which states that sample means of a sufficiently large number of samples containing an uniform number of variables, extracted from the same population, yield a normal or close to normal distribution in spite of the randomness of this sampling.

Closed population See *fixed population*.

Cohort Population whose individuals share common characteristics.

Confidence interval A range of values of a sample which, expectedly, contain the parameters of the general population where it was taken from, under a conventional degree of probability (confidence level).

Confounder "Concealed" independent variable which, along with the independent variable, co-influences the dependent variable, then "confounding" the investigator. For example, there is a report of an elevated prevalence of high body weight (dependent variable) in a cohort of teachers from a local school. The investigator verifies that a certain brand of sweetener (independent variable) is added to the coffee on a daily basis. He or she then associates the high body weight findings to this brand of sweetener. Nevertheless, one eventually finds out that the coffee is inadvertently prepared with a high content of sugar ("concealed" independent variable) the confounder , the element which in fact is causing the teachers' body weight to increase.

Control group Collection of individuals used as a reference parameter for the so-called experimental group, against which the latter is tested.

Cost The sum of microcosts (see *microcost*) of a diagnosis-related group.

Covariate In observational studies context, a covariate is a variable different from the major variables—condition or exposure—, but which can influence the outcome as well.

Critical value Conventional cutoff value which separates statistical significance from statistical nonsignificance in a study. It corresponds to α (see α).

Degree of freedom In statistics context, degree of freedom (ν) is a measure of available numerical possibilities of a variable in a set of variables. For example,

• $2 + 1 + 3 = 6$—This equation has THREE degrees of freedom—its first, second, and third elements—for they were "free" to vary in order to sum 6.

• $2 + 2 + z = 6$—This equation has TWO degrees of freedom—its first and second elements—for they were "free" to vary. However, for this equation to sum 6, the third element—z—can NOT be "free" to be any number other than 2.

Dispersal The degree a set of observations deviate from their mean.

Drop-out A subject who withdraws from a study for any reason.

Dynamic population Open population. In a general context, all of its members have a changeable status (e.g., people who live in the eastern coast). In epidemiological, observational or intervention study contexts, the term expresses a type of population which presents a certain characteristic susceptible to change along with the study (e.g., high arterial blood pressure patients treated with diuretics).

Effect size The difference between the results of two groups.

Endpoint Efficacy variable chosen as the parameter meant to determine the outcome of a clinical trial. Endpoints can be subdivided into primary endpoint and secondary endpoint.

Estimator Sample statistic meant to estimate its related parameter (see *parameter*). For example, $\bar{x}$—sample arithmetic mean (the estimator)—used to estimate μ—population arithmetic mean—of the corresponding population.

Evidence-based medicine Accurate use of updated and systematically harvested literary evidence, meant to support bedside decision-making processes.

False–negative Situation in which a diagnostic test indicates absence of a condition in a patient who actually has it.

False–positive Situation in which a diagnostic test indicates presence of a condition in a patient who actually does not have it.

f_h (fraction$_{harm}$) See *fraction$_{harm}$*.

Fixed population Closed population. In a general context, all of its members have a lifelong status (e.g., WWII veterans). In epidemiological, observational, or intervention study contexts, the term expresses a type of population which presents a certain characteristic supposed to last throughout the study (e.g., patients who fulfilled a food questionnaire).

fraction$_{harm}$ (f_h) Risk of a control-patient in presenting an adverse reaction, relatively to a treated patient.

fraction$_{treatment}$ (f_t) Risk of a control-patient in presenting an undesired event, relatively to a treated patient.

f_t (fraction$_{treatment}$) See *fraction$_{treatment}$*.

Gray literature Hard-to-explore bibliographic sources such as postgraduation dissertation and theses, congress abstracts, official reports and studies in non-English languages.

Historical (literature) controls Patients treated in the past with a standard drug or vaccine, who can be used as control group (see *control group*) in a current study.

Interaction In multiple linear regression context, interaction corresponds to the state where an independent variable indirectly influences a dependent variable, through *interaction* with a second independent variable. For example, suppose there is a high incidence of respiratory diseases (dependent variable) in a cohort of office employees, apparently due to a low-temperature environment (independent variable) provided by air conditioning. Nevertheless, we verify that the air conditioning system is contaminated by a species of fungus which grows in low temperatures (second independent variable). Hence, independent variable (low-temperature environment) and second independent variable (cryophile fungus) interact. This definition can be extended to contexts different from multivariable analysis.

Intervention study Type of analytical study where the investigator proactively influences the design and course of the investigation, with the general goal of demonstrating the difference—generally as a favorable outcome—between two groups. The term "clinical trial" is frequently used interchangeably.

Kurtosis The degree of sharpness or bluntness of a distribution curve in a graph.

Lethality The intrinsic capacity of causing death (lit.).

Literature controls See *historical controls*.

Microcost The sum of all treatment cost items of an individual patient.

Morbidity The quality or state of being unhealthy (lit.).

Mortality The quality or state of being mortal (lit.).

n The number of individuals in a sample.

N The number of individuals in a population.

Null hypothesis The no–difference presumption regarding an investigator's hypothesis. It is meant to be statistically tested against the alternative hypothesis (H_1) which presumes otherwise.

Observation In the context of this book, it corresponds to a discrete individual or phenomenon counted in a sample or population.

Odds The ratio of the probabilities of the two possible states of a binary variable. For example, the probability of symptomatic remission against the probability of symptomatic worsening.

Open population See *dynamic population* in population.

Outlier An observation that markedly deviates from the average of a set of variables in a population or sample.

Parameter In the setting of population estimation based on sampling, it corresponds to a population variable estimated by its corresponding estimator (see *estimator*). For example, $\bar{x}$ (mean) of a given sample is used to estimate μ (mean)—the parameter—of the population.

Patient expected event rate (PEER) Expresses the proportion of patients not exposed to the putative harming agent but who nevertheless present the harm. It may be inferred from literary sources consistent with the potential harming agent.

PEER See *patient expected event rate*.

Person-time Measure that combines people and the timespan in which these people are subject to an event. It is determined by the formula:

$$\mathrm{PT} = n_p \cdot \Delta t$$

Where PT = person-time, n_p = number of people, and Δt = timespan.

If one person was exposed for 4 years, we must consider these 4 years as an unit by multiplying that one person by 4 years (4 persons-year). Saying differently, following 1000 persons annually along 5.5 years would make 5500 person-years.

Population In this book this term has two meanings, according to the context: (1) general and broader universe of individuals from where samples (see *sample*) are taken or (2) the total of individuals occupying an area or making up a whole.

Probability The quantitative expression of the chance that an event might occur.

Random error A study anomaly associated with a randomly erroneous parameter or process. Unlike a systematic error (see *systematic error*), the effect generated by random error can be corrected by increasing n, because in the end the error will be evenly spread. For example, in a study on the efficacy of a thrombolytic drug for the treatment of acute ischemic stroke, head MRIs are distributed between equally experienced radiologists A and B, for diagnosis. It is expected that an eventual diagnostic error from radiologist A might be compensated by radiologist B and vice versa. It is also expected that increasing n would promote the balance between both even further.

Rate Dynamic ratio between two distinct quantities (e.g., number of births per year).

Ratio Figure obtained by dividing one number by another, without necessarily implying a correlation between them. For example, a population of 10 adults and 5 children implies a ratio of 10:5 or two adults for each child.

Sample A subset of individuals selected from a population (see *population*).

Sample size See *n*.

Sampling distribution The probabilistic distribution of a statistic or a group of statistics (see *statistic*). For example, we have a population of 1000 values of a variable, from which 100 samples of 10 variables each are randomly taken, each one with its corresponding statistic. The resulting 100 statistics collection represents the sampling distribution.

Sensitivity analysis An analytical resource in which a given situation is artificially taken to the limit in order to test the robustness of the applied model.

Skewness The degree of unilateral inclination of a distribution curve in a graph.

Statistic A variable which represents a certain statistical aspect of a sample (e.g., mean and standard deviation). Important: statistic (or its plural form) should NOT be confused with the term Statistics (see *Statistics*).

Statistics Science concerned with the organization, description, analysis and interpretation of experimental data.

Subgroup A fraction of a group (see *group*).

Surrogate A less accurate type of parameter, nevertheless circumstantially more convenient, which has the potential for replacing the ideal parameter in a study. For example, serum glucose levels rather than serum osmolarity for determining the efficacy of a new insulin formulation in the treatment of hyperosmolar diabetic coma (the facility is not provided with serum osmolarity test but would be equipped for blood glucose test).

Systematic error Study anomaly associated with an intrinsically erroneous parameter or process. Differently from random error (see *random error*), bias generated by systematic error cannot be corrected by increasing n, for the error will multiply itself along with n. For example, in a study on the efficacy of a thrombolytic drug for the treatment of acute ischemic stroke, head MRIs are distributed between an experienced radiologist and an unexperienced radiologist, for diagnosis. The difference between the radiologists' expertise levels might generate bias, not expected to be rectified by n increase.

Target population Population (see *population*) represented by a specific characteristic of interest. For example, pediatric patients with influenza virus infection.

Type I error Spurious rejection of the null hypothesis (see *null hypothesis*). For example, we state that a vasodilator IS efficacious in the prevention of angina pectoris episodes, when actually it is NOT (null hypothesis—the vasodilator is NOT efficacious—was erroneously rejected).

Type II error Spurious acceptance of the null hypothesis (see *null hypothesis*). For example, we state that a vasodilator is NOT efficacious in the prevention of angina pectoris episodes, when actually it IS (null hypothesis—the vasodilator is NOT efficacious—was erroneously accepted).

Variable Quantifiable parameter that may assume any one of a set of values. For example, body temperature, arterial blood pressure, serum thyroxin levels, etc.

Washout A time interval introduced between two study phases, in order to minimize carryover effects from the former to the next.

Weighted mean An average determined through the sum of a series of different elements, that takes their relative importance into account.

Index

Note: Page numbers followed by "*f*" and "*t*" refer to figures and tables, respectively.

Printed in the United States
by Baker & Taylor Publisher Services